Alan Dél Carlos Gomes Chaves
Aline Costa Ferreira

Evaluation of water for irrigating vegetables in the Paraiba Hinterland

Alan Dél Carlos Gomes Chaves
Aline Costa Ferreira

Evaluation of water for irrigating vegetables in the Paraiba Hinterland

Use of water for irrigating vegetables in the sertão of Paraíba state

ScienciaScripts

Imprint

Any brand names and product names mentioned in this book are subject to trademark, brand or patent protection and are trademarks or registered trademarks of their respective holders. The use of brand names, product names, common names, trade names, product descriptions etc. even without a particular marking in this work is in no way to be construed to mean that such names may be regarded as unrestricted in respect of trademark and brand protection legislation and could thus be used by anyone.

Cover image: www.ingimage.com

This book is a translation from the original published under ISBN 978-613-9-64886-3.

Publisher:
Sciencia Scripts
is a trademark of
Dodo Books Indian Ocean Ltd. and OmniScriptum S.R.L publishing group

120 High Road, East Finchley, London, N2 9ED, United Kingdom
Str. Armeneasca 28/1, office 1, Chisinau MD-2012, Republic of Moldova, Europe
Printed at: see last page
ISBN: 978-620-7-79818-6

SUMMARY

The intensive agriculture developed has different environmental impacts on water quality. It is therefore necessary to monitor various quality indicators, including the evaluation of pesticide residues. The aim of this work was to study the quality of groundwater used for irrigating vegetable production in the community of Várzea Comprida dos Oliveiras in the municipality of Pombal, PB, considering its use for irrigation. The study was carried out in the municipality of Pombal, PB, following visits to the Várzea Comprida dos Oliveiras and Bezerro communities, which produce vegetables irrigated with groundwater from tube wells in the region. 19 water samples were collected and analysed on site and sent to the soil and water analysis laboratory (IFPB/Sousa). There was a reduction in the electrical conductivity of the water from February onwards, a period marked by the start of the rains, demonstrating the effect of rainwater in diluting the salts. The Piranhas River showed higher turbidity values from February onwards, as the rains during this period are able to stir up the material deposited at the bottom of the river and wash away that which is in the riverbed. Of the two salts studied, bicarbonate was the ion that showed the highest values in the water, with wells 4, 6 and 10 showing 11.3, 10.34 and 10.42 mmolc L^{-1} . Chloride and sodium showed results ranging from 0.4 to 2.7 mmol$_c$ L^{-1} and 0.2 and 3.68 mmol$_c$ L^{-1} respectively, not causing concern for use in irrigation, taking into account efficient irrigation management. The sodium adsorption ratio showed the highest values in wells 4 (2.67 mg L^{-1}) and 13 (2.6 mg L)$^{-1}$

Key words: Water quality, wells, monitoring.

Summary

CHAPTER 1

INTRODUCTION

Among the fundamental natural resources, water is the most important, as its availability and access are necessary for all kinds of life on the planet, as well as for most means of production (SARDINHA et al., 2008).

The three main categories of water use are agricultural, industrial and domestic, with the agricultural sector standing out as the biggest user in most developing countries. It is estimated that irrigation in these countries uses 70 per cent of all the water withdrawn from rivers, lakes and underground springs (PRUSKI et al., 2004).

The intensive agriculture developed has different environmental impacts on water quality. It is therefore necessary to monitor various quality indicators, including the evaluation of pesticide residues.

Most of the chemical contaminants present in groundwater and surface water are related to industrial and agricultural sources. The variety is enormous, especially pesticides, volatile organic compounds and metals (HU; KIM, 1994).

The search for greater productivity in agriculture has led to the inappropriate consumption of chemical products, such as fertilisers and pesticides, implying serious problems for the quality and quantity of surface water (TELLES, 2002; MILHOME et al., 2009). In groundwater, these compounds are considered a potential threat to the quality of this source, especially when the aquifers are located in or near a region used for agricultural activities (SÁ BARRETO, 2006).

Pesticides and related products are products and components of different processes and are used in production, storage and processing in agriculture, pasture, forest protection and other environments, to preserve them from the harmful action of harmful beings, as well as substances and products used as defoliants, desiccants, growth stimulators and inhibitors (BRASIL, 1989).

Pesticides stand out as contaminants due to their intensity, which can pose environmental risks and serious health problems. Their presence in water sources can make it difficult to treat the water due to the possible need for more complex technologies than those normally used for potabilisation. Therefore, the use and consequent exposure to agrochemicals can jeopardise the multiple uses of water by the present community, especially in neighbouring farming areas.

The consumption of vegetables is fundamental in any nutritionally adequate menu, due to their vitamin, mineral and fibre content, low calorie intake and because they increase food residue in the gastrointestinal tract (NASCIMENTO et al., 2005). Lately, there has been a change in the population's diet in terms of increased consumption of fresh vegetables, as these foods provide numerous benefits to the body, such as the development and organic regulation of the body (OLIVEIRA et al., 2006).

However, vegetables eaten raw are one of the important food groups responsible for transmitting enteric diseases. Contamination can occur in the garden, resulting from the use of inadequate irrigation water or fertilisers, during transport or through handling at the point of sale; and successive handling increases the chances of contamination (TAKAYANAGUI, 2001).

In this context, this work seeks to assess the contamination of vegetables sold in the Várzea Comprida dos Oliveiras community in the city of Pombal-PB, considering their use for domestic supply and irrigation, in order to provide information that can serve to prevent and reduce the risks of contamination in the places surveyed.

CHAPTER 2

LITERATURE REVIEW

2.1. Water

World agriculture needs water in quantity and quality to produce food (SOUTO, 2005). The water used for irrigation comes from rivers, streams and lakes close to the gardens, and almost no public water supply is used, as the demand for irrigation is high and the cost is high. It is transported, without any prior treatment, via canals or pumps, from the river or stream to the gardens (OLIVEIRA; GERMANO, 1992). And it can present biological contaminants such as faecal coliforms when associated with domestic sewage discharges or even the presence of animals near these areas (SOUTO, 2005).

Therefore, food that comes into contact with contaminated water and is eaten raw is a likely source of microorganisms (PACHECO et al. 2002). The main biological agents found in these waters are pathogenic bacteria, viruses and parasites. These organisms found in water or food represent one of the main sources of morbidity in the country and are responsible for countless cases of infectious and parasitic diseases, enteritis and diarrhoea in children, which can even lead to death (ALMEIDA FILHO, 2008). Sanitary control of the water used in agricultural practices is important for maintaining the health of the population (MORETTI, 2003).

According to Moretti (2003), the following aspects should be considered to avoid contamination of water sources used in vegetable production:

- Identification of water supply sources;
- Observe the presence of livestock in the vicinity of the water source used;
- Systematically prevent wild animals and unauthorised persons from approaching water sources;
- Avoid storing organic manure near water sources;
- Have a maintenance schedule for the water storage tanks;
- Carry out regular tests on the quality of the water used.

2.2 Water quality

Agro-industry contributes to water pollution/contamination with untreated organic waste generated and chemical products used (pesticides, fertilisers) whose residues seep into the soil or are carried by rain to surface water sources (ESTEVES, 1988).

Monitoring water quality parameters is a basic tool for assessing environmental changes caused by anthropogenic action (Molozziet al., 2006).

In order to maintain water quality control, it is necessary to monitor indicators, including the evaluation of pesticide residues, most of which reach underground and surface water bodies through leaching, contaminating and jeopardising water quality for other purposes such as fish farming, irrigation, human consumption and others.

The World Health Organisation (WHO) Guides have been used as a global reference in guiding national standards and legislation on the quality of water for human consumption. They have usually had a major influence on the drafting and updating of Brazilian legislation on drinking water quality (HELLER et al., 2005).

The environmental impacts on water resources generated by agricultural activities cannot be dissociated from the impacts on the production areas themselves, and their monitoring and preventive measures must always be integrated in a systemic way (ANDREOLI, 1993).

In Brazil, the National Environment Council (CONAMA) in its Resolution No. 357, of March 2005, establishes quality levels for environmental waters, assessed by specific parameters and indicators, in order to ensure the use of fresh, saline and brackish waters (BRASIL, 2005). CONAMA Resolution No. 396 of April 2008 establishes quality levels for groundwater, taking into account the predominant uses (BRASIL, 2008).

Water quality standards are used to regulate the quality levels to be maintained in a body of water, in accordance with its intended use. The use of quality standards serves two purposes: to maintain the quality of the watercourse or to define the target to be achieved; the basis for defining the levels of treatment to be adopted in the basin, so that the effluents discharged do not alter the characteristics of the watercourse established by the standard (PORTO et al., 1991).

2.3 Pollution and contamination of water sources

Federal Law 6.938, of 31 August 1981, which provides for the National Environmental Policy, defines pollution as the degradation of environmental quality resulting from activities that, directly or indirectly,

a) jeopardise the health, safety and well-being of the population;

b) create adverse conditions for social and economic activities;

c) unfavourably affect biota;

d) affect the aesthetic or sanitary conditions of the environment;

e) release materials or energy in disagreement with established environmental standards.

Pollution of water bodies occurs in point and diffuse form and from natural or anthropogenic sources (LIBÂNIO, 2005). Point source pollution is when the pollutant reaches the body of water in a concentrated way in space; in diffuse sources of pollution, the pollutants enter the body of water distributed along its length (VON SPERLING, 2005).

The various chemical contaminants present in groundwater and surface water can be portrayed in terms of concentrations and variations in physical, chemical and biological parameters (VON SPERLING, 2005).

The selection of parameters to be sampled must be associated with the characteristics of the site, water uses and quality objectives. In this sense, parameters that provide, among other things, the following essential information should be assessed: temperature, pH, colour, turbidity, electrical conductivity, dissolved oxygen, biochemical oxygen demand, suspended solids, orthophosphate, ammoniacal nitrogen, nitrate, chlorides, oils and grease, phenols, arsenic, cadmium, chromium, iron, manganese, mercury, phytoplankton and chlorophyll A (VON SPERLING, 2001).

In the European Union, the limits for drinking water are stricter, where a maximum concentration of 0.1 µg L-l is permitted for any individual pesticide and 0.5 µg L-l for the total of pesticides present in drinking water. For surface water, the maximum permitted limit is in the order of 1 to 3 µg L-l (EUROPEAN COUNCIL, 1998).

Of the 22 pesticides covered by Ordinance 518/2004, the majority are herbicides. In terms of toxicological classification, substances in Class III (moderately toxic) predominate, followed by Class I (extremely toxic). Of these substances, several do not have authorisation for use in Brazil, so the monograph
The active ingredient is not available from the National Health Surveillance Agency (ANVISA).

2.4. Groundwater

Groundwater is all the water that lies below the earth's surface and is closely linked to surface water, resulting from the slow process of rainwater infiltration that fills the pores and interstices of the soil, forming aquifers (LIBÂNIO, 2005).

Groundwater can be collected through shallow or deep wells, infiltration galleries or by tapping springs (BRASIL, 2006). In rural areas, shallow wells and springs are the main sources of water supply and are highly susceptible to contamination (RIGOBELO et al, 2009). In most cases, this problem is due to the lack of sewage collection networks, leading to the use of black pits and the inadequate digging and lining of wells (LIBÂNIO, 2005).

Shallow wells, also known as manual or phreatic wells, are manual or mechanical excavations with a cylindrical section and a variable diameter (a few centimetres to metres) whose depths are defined by the water levels of the respective aquifers. Due to their shallow depths and the nature of the areas where they are built, shallow wells contribute to pollution of the groundwater aquifer

(HELLER and PÁDUA, 2006). This type of well is divided into three types:

• Simple hand-dug wells: these are vertical excavations carried out using hand tools; they generally have circular sections and a diameter of around one metre.

• Shallow tube wells: vertical excavations made by auger or metal rods, usually in unconsolidated material;

• Amazon wells: vertical excavations where water is produced and stored at the same time. They are shallow wells, generally up to 10 metres deep and between 3 and 6 metres in diameter.

Article 22 of MS Ordinance 518/2004, which lays down water potability standards in Brazil, states that all water supplied collectively must be subjected to a disinfection process to guarantee compliance with the microbiological standard. In this respect, chlorine, chlorine dioxide, ozone and ultraviolet radiation are among the most commonly used disinfectants (BRASIL, 2006).

2.5 Pesticides

Pesticides, pesticides, pesticides, pesticides and biocides are different names given to chemical substances or mixtures of substances, natural or synthetic, whose purpose is to kill, control or combat, in some way, different pests (mites, rodents, weeds, bacteria and other forms of animal or plant life) that attack, damage or transmit diseases to plants, animals and humans (SANCHES et al., 2003).

Pesticides can be classified according to the type of pest they control, the chemical group they belong to, their mode of action and their effects on health and the environment (MENEZES, 2006). Depending on the type of pest, the classification is made in relation to the group of target organisms involved in the use of pesticides. These can be grouped into several classes: insecticides (used to control insects), fungicides (used to control fungi), herbicides (used to control invasive plants), defoliants (used to control unwanted leaves), fumigants (used to control soil bacteria), rodenticides or rodenticides (used to control rodents/rats), nematicides (used to control nematodes) and acaricides (used to combat mites) (RIBAS; MATSUMURA, 2009).

In the case of the intensive and indiscriminate use of pesticides, this activity is a growing problem that contributes to environmental pollution. The contamination of surface and groundwater sources and, above all, their occurrence in water for human consumption has been the subject of constant concern (MENEZES, 2006).

Brazil is considered the third largest agricultural exporter in the world, behind only the United States and the European Union (IBAMA, 2010). In 2008, Brazil became the world's largest consumer of pesticides. °According to a survey carried out by the National Union of the Agricultural Defence Products Industry (SINDAG), sales of pesticides totalled US$ 7.125 billion, followed by the United

States (US$ 6.6 billion) (ANDEF, 2009).

Also in relation to target organisms, pesticides can be: contact, when the target organism is directly hit by the substance; ingestion, when the substance acts after being ingested by the target organism; or systemic, when the substance is absorbed by a part of the plant and translocated to all the vascular tissues (MARASCHIN, 2003).

In terms of toxicological categorisation, ANVISA classifies pesticides into four different classes associated with a colour band on the product label. The class
Class I (red label) covers compounds considered highly toxic to humans; Class II (yellow label), those that are moderately toxic; Class III (blue label), those that are slightly toxic; and Class IV (green label), compounds considered practically non-toxic to humans (SANCHES et al., 2003).

When pesticides are applied to the soil, they can remain in their original form or undergo chemical or biological transformation processes. In this sense, the reduction of its concentration to 50% of the initial mass applied to the soil is expressed as the half-life (DT50) of the product in the soil (OLIVEIRA, 2007).

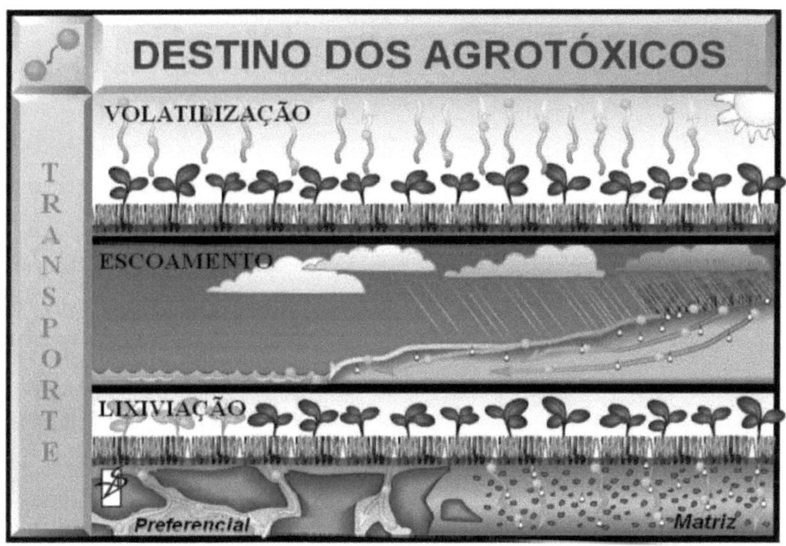

Figure 1- Transport and destination of pesticides in the environment.
Source: Adapted from BLESSING (2001).

The occurrence of heavy rainfall, especially after the application of pesticides, can cause an increase in the concentrations of these molecules in water and the further away the episodes of rainfall, the lower the potential for leaching and run-off of these chemicals (DEUBERT, 1990; MENEZES, 2006).

2.6 Pesticides vs. Brazilian legislation

According to PASCHOAL (1979), the first products used in the country to control pests were of mineral and botanical origin. The first organic insecticide

- The first synthetic product to be used was DDT, introduced in Brazil at the end of 1943 under the name Gesarol. From 1946-47 onwards, other products such as BHC and ethyl parathion were introduced and used on crops.

Brazil has become the world's third largest consumer of agrochemicals, with a very precarious institutional structure in terms of legislation, research, inspection, commercialisation, training, professional ethics and rural extension (PASCHOAL, 1983).

The use of pesticides in Brazil was regulated by Federal Law[0] 7.802 of 11 July 1989. Known as the "Law on Pesticides", this law, regulated by Decree 4.074 of 4 January 2002, governs: research, experimentation, production, packaging and labelling, transport, storage, marketing, commercial advertising, use, import, export, the final destination of waste and packaging, registration, classification, control, inspection and supervision of pesticides, their components and the like (BRASIL, 1989).

The current drinking water standard for chemical substances that pose a health risk is explained according to type (inorganic, organic, pesticides, disinfectants and secondary products of disinfection). Due to their importance in terms of use in the country and health risks, agrochemicals (27 in total) were considered separately.

Pesticides, pesticides, pesticides, pesticides and biocides are various names given to chemical substances or mixtures of substances, natural or synthetic, whose purpose is to kill, control or combat, in some way, the various pests (mites, rodents, weeds, bacteria and other forms of animal or plant life) that attack, damage or transmit diseases to plants, animals and humans (SANCHES et al., 2003).

Law 9.974, of 6 June 2000), regulated by Decree 4.074 of 4 January 2002, are defined as products and/or components of physical, chemical or biological processes intended for use in the production, storage and processing of agricultural products, in pastures, in the protection of native or established forests, in other ecosystems and in urban, water and industrial environments, whose purpose is to alter the composition of flora and fauna in order to preserve it from the harmful action of living beings considered harmful, as well as substances and products used as defoliants, desiccants, growth stimulators and inhibitors (BRASIL, 2002).

In Brazil, the process of registering pesticides involves three bodies: The Ministry of Health (MS), the Ministry of Agriculture, Livestock and Supply (MAPA) and the Ministry of the Environment, through the Brazilian Institute of the Environment and Renewable Natural Resources

(IBAMA). The Ministry of Health, through the National Health Surveillance Agency (ANVISA), is responsible for the toxicological evaluation and classification of pesticides and, together with MAPA, is responsible for monitoring pesticide and related residues in products of plant origin. IBAMA/MMA is responsible for assessing and classifying the potential for environmental danger (BRASIL, 2002a).

Living with pesticides is part of the daily routine of both rural and urban populations, who often unconsciously consume food and water containing residues of these products (MEDEIROS, 1988).

2.7. Irrigation

According to Mantovani (2000), existing irrigation systems include conventional sprinkler irrigation, hydraulic cannon irrigation, self-propelled irrigation, centre pivot irrigation and drip irrigation (localised irrigation). The author also states that the choice of any of these systems will depend on various factors such as: type of soil, topography and size of the area, climatic factors, factors related to crop management, water deficit, the producer's investment capacity and the cost of the irrigation system.

In sprinkler irrigation (Figure 2), the water is directed at the plants through a pressurised mechanism that causes the water jet to split into small drops before being distributed over the ground. The mechanism that causes the water jet to be fractionated is called a sprinkler, and there are different characteristics and models depending on the manufacturer and the purpose of the equipment (FERREIRA, 2005).

The hydraulic cannon system (Figure 3) is a variation of the conventional sprinkler system in which the modification is the use of larger sprinklers, which allow for greater spacing between lines and sprinklers, thus reducing labour and allowing larger areas to be irrigated (MONTOVANI, 2002).

Figure 2: Irrigation system **Source:** Google, 2014
Figure 3: Sprinkler system **Source:** author, 2014.

Localised irrigation, known internationally as micro-irrigation, is the application of water to the soil in a range restricted to the root system. In this way, only the surface of the soil is wetted, consequently reducing the direct evaporation of water from the soil into the atmosphere and allowing for greater efficiency in the application and control of the water applied when compared to sprinkler and surface irrigation systems (RODRIGO LÓPEZ et al., 1992).

According to Silva et al. (2003), the most widespread localised systems are drip (Figure 4) and micro-sprinkler (Figure 5).

Figure 4: Drip irrigation **Source:** Google, 2014
Figure 5: Micro-sprinkler irrigation **Source:** author, 2014.

The drippers are connected to the lateral lines, capable of dissipating the pressure available in the lateral line and applying small, constant flows. The flow rate generally varies between 1 and 20L.h-l, applied drop by drop, under service pressures varying between 5 and 25 mca (BERNARDO et al., 2005).

According to Amaral (apud PIZARRO CABELLO, 1990) micro-sprinklers are small plastic sprinklers, usually installed on the lateral line, with flow rates applied in a sprayed form in the range of 20 to 150 L.h-1, under service pressures varying from 10 to 20 mca. It is a type of emitter that offers more advantages to crops with wider spacing and a large root system (bananas, lemons, mangoes, etc.), as it has a larger wetted radius and higher flow rates than drippers (Amaral apud LOPES, 2006).

In addition to drippers and micro-sprinklers, there are alternative emitters such as the "Microjet" or "Microspray" (Figure 9), which work in an intermediate range between drippers and micro-sprinklers in terms of pressure and flow (BUFON et al., 2000; SOUZA, 2000).

Figure 6: Localised system with "Microjet" type emitter.
Source: Google, 2014.

According to Amaral (apud Silva et al.2003), among the many benefits provided by localised systems:

- Control the amount of water to be supplied to the plant;
- Easy distribution of fertilisers and other chemical products near irrigation water;
- Low energy consumption;
- High potential water application efficiency;
- Low labour costs and easy automation;
- Less weed growth between planting rows.

As disadvantages to this type of equipment, Lopes (2006) cites its high implementation cost and sensitivity to clogging. Also noteworthy as disadvantages of this system are leaks resulting from damage caused by animals and poor cultivation, the need for maintenance and monitoring (SKAGGS, 2001).

The classification guidelines proposed by the FAO (AYERS & WESTCOT, 1999) are also highly recommended, especially when there are high levels of salts.

Table 1. Guidelines for interpreting water quality for irrigation[1] , according to Ayres and Westcot (1999).
Source: Ayres and Westcot, (1999).

Potential Irrigation Water Problem

	Unit	None	Degree of Restriction of Use		Severo
			Low to moderate		
Salinity					
ECa[2]	dS m^{-1}	<0,7	0,7	3	>3,0
SDT[3]	mgL^{-1}	<450	450	- 2000	>2000
Infiltration					
RAS[4]	0 - 3and ECa	>0,7	0,7	-0,2	<0,2
	3 - 6	>1,2	1,2	-0,3	<0,3

6 - 12	>1,9	1,9	-0,5	<0,5
12 - 20	>2,9	2,9	1,3	<1,3
20 - 40	>5,0	5	-2,9	<2,9
Toxicity of specific ions				
Sodium (Na)[3]				
Surface irrigation RAS	<3	3	9	>9
Sprinkler irrigation meq L^{-1}	<3	>3		
Chloride (Cl)[3]				
Surface irrigation meq L^{-1}	<4	4	10	> 10
Sprinkler irrigation meq L^{-1}	<3	>3		
(Other ions that affect sensitive crops)				
Nitrogen (N - (NO)$_{36}$ mgL^{-1}	<5,0	5	-30	>30
Bicarbonate (HCO$_3$) (conventional sprinkling only) meq L^{-1}	<1,5	1,5 -8 ,5		>8,5
Ph		Standard range 1: 6,5 - 8,4		

Adapted from UNIVERSITY OF CALIFORNIA COMMITE OF CONSULTANTS (1974)

ECa = Electrical conductivity of water, in dS m' at 25°C

[3]TDS = Total Dissolved Solids (mg. L')[1]

[4]RAS = Sodium Adsorption Ratio, sometimes referred to as Rna. For a given RAS value, the infiltration rate increases as salinity increases. The potential infiltration problem is assessed using the RAS and the ECa.

[5] Most tree crops and woody plants are sensitive to sodium and chloride; in the case of surface irrigation, the indicated values are used.

[6]It means nitrogen in the form of nitrate, expressed in terms of elemental nitrogen.

2.8 Location of the study area

The study area is located in the Mesoregion of Sertão and Homogeneous Microregion of Sousa, in the state of Paraíba, situated in the north-eastern region of Brazil (Figure 7). The municipality is one of the oldest in the state of Paraíba, and the second largest in the state of Paraíba in terms of land area, with 889km^2 . According to the IBGE (Brazilian Institute of Geography and Statistics), the population in 2010 was estimated at 32,443 inhabitants.

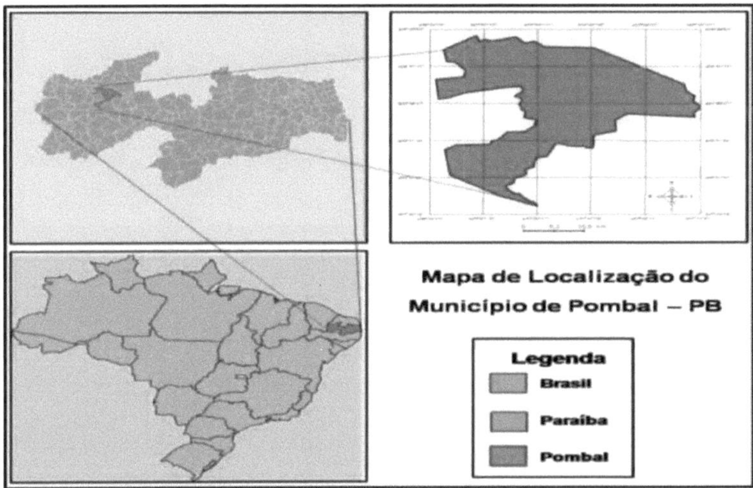

Figure 7. Location of the study municipality.

Source: Google, 2014

The study will be carried out in the Várzea Comprida dos Oliveiras community, located in a rural area 11 kilometres from the municipality of Pombal - PB. The community is located in the Sertão Paraibano mesoregion and has the following geographical coordinates: Latitude 6° 45' 23" S and longitude 37° 51' 49" O. Google Earth (2012).

The research area was chosen because the community has great potential for producing vegetables, and because it maintains a way of life geared towards the use of natural resources and organic vegetable cultivation practices, as well as because of its history and the references of its residents, and the ease of access to it.

2.9 Piranhas-Açu catchment area

The federally-owned Piranhas-Açu is the main river in the basin, as it rises in the municipality of Bonito de Santa Fé, in the state of Paraíba, and follows its natural course through the state of Rio Grande do Norte, flowing into the Atlantic Ocean on the Potiguar Coast. It is an important basin for the states of Rio Grande do Norte and Paraíba, as it is here that the Armando Ribeiro Gonçalves dam and the Coremas - Mãe D'Agua reservoir system are located, as well as the Várzea Comprida and Bezerro communities, the site of the research study.

Figure 8: Piranhas-Açu River Basin
Source: AESA, 2014

The Piranhas-Açu Hydrographic Basin, totally inserted in the semi-arid climate of the northeast, has a total drainage area of 43,681.50 Km2, of which 26,183.00 Km2, corresponding to 60% of the area in the state of Paraíba, and 17,498.50 Km2, corresponding to 40% of the area in the state of Rio Grande do Norte. It has a total population of 1,363,802 inhabitants, of which 914,343 inhabitants (67%) are in the state of Paraíba and 449,459 inhabitants (33%) in the state of Rio Grande do Norte.

2.10 Sustainable development as an element of local coexistence.

2.10.1 Sustainable development

Sustainable development is understood as: "development that meets the needs of the present without compromising the ability of future generations to meet their own needs" (CMMAD, 1991, p. 09). This is the most widespread concept of sustainable development, created by the World Commission on Environment and Development, also known as the Brundtland Commission, which held forums for debate between 1983 and 1987. The Commission was set up by the United Nations to assess the planet's main environmental and development problems and formulate possible proposals to solve them, with the intention of ensuring that human progress is sustainable for future generations in their development process.

In this way, the contribution of this research is to help in human and social promotion, with a view to sustainability in its main dimensions: social, economic and environmental.

It should be emphasised that family farming can contribute as an element of sustainable development, as it is a production activity that generates income, requires time for dedication and productive activities, enables the involvement of family members and collective actions in the local production process.

CHAPTER 3

MATERIALS AND METHODS

3.1.General aspects

This fieldwork was carried out in the municipality of Pombal, PB, following visits to the Várzea Comprida dos Oliveiras and Bezerro communities, which produce vegetables irrigated with groundwater from tube wells in the region. 19 water samples were collected and analysed on site and sent to the soil and water analysis laboratory (IFPB/Sousa).

The work was carried out in four stages, the first being a reconnaissance of the study area. In the second stage, the producers of the pessoahnente community were recognised; the existing artesian wells in the Várzea Comprida dos Oliveiras community and the Bezerro community were surveyed; the agricultural crops grown in the region and the pesticides used were surveyed. The third stage involved georeferencing and defining the tube wells, and the fourth stage involved monitoring the groundwater in the wells studied.

3.8.1. Step 1

3.8.1.1. Area reconnaissance

To begin this work, it was necessary to recognise the community under study through field visits. Initially, residents were interviewed verbally to find out about the community's producers, the predominant crops grown in the region, the most commonly used pesticides and the difficulties faced by these farmers with regard to agricultural activities. In this first stage, the properties were visited and the producers informed about the work to be carried out.

3.8.2. Stage 2

3.8.2.1. Identification of producers and survey of existing wells in the locality

At this stage, the first direct contact was made with the producers based on identification, where the criterion for choosing the informants was random, regardless of gender, but a representative who was willing to provide the necessary information for the work and who is involved with the project in progress, thus obtaining the following information: most frequent use of the water available on the property; existence of treatment of the water used for consumption, number of people living on the site; type of sewage treatment; crops produced; use of pesticides; access to the purchase of pesticides and technical guidance on care when applying the products; complaints regarding the organoleptic aspect of the water (colour, odour and taste) in relation to the application of pesticides and alteration of the quality of surface and underground water and a survey of the crops grown on each property.

3.8.3. Step 3

3.8.3.1. Choice of monitoring points

The monitoring points for the groundwater wells were defined based on the results of information previously provided by farmers in the region, and based on their proximity to agricultural areas.

Each monitoring to be carried out in the areas selected for this study was carried out:

- Environmental conditions: measuring the water level of the wells monitored in order to relate the quantity of water to the level of water in the well; assessing the uses of the water on the property; the rainfall at the time of collection, etc;
- Agronomic conditions: crops grown; area cultivated; pesticides applied, how and when they were applied, whether any pesticides were applied at the time of monitoring, type of irrigation system, etc.

3.8.4. Step 4

3.8.4.1. Groundwater monitoring

Some wells were selected for water collection. A map was drawn up of the local community containing the georeferencing of the wells and their location.

The groundwater samples were collected during the dry season, from December 2013 to April 2014, at the outlets of the artesian well pumps after initial disposal. If there was no water coming from the pumps, it was collected directly from households with piped groundwater, allowing the water to drain for a few minutes before collection. Water was collected manually from the wells.

Before storing the samples, the containers were filled three times with the sample itself and then discarded.

Figure 9: Tube well: pump outlet.

Source: Author, 2014.

3.2 Physico-chemical analyses

All collection procedures, preparation of collection bottles, transport and packaging of samples for physico-chemical analyses will follow the recommendations of the Standard Methods for the Examination of Water and Wastewater 20TH ed. (APHA, 1998).

The samples will be packed in Styrofoam boxes and kept on ice until they reach the Soil and Water Analysis Laboratory at the Federal Institute of Education, Science and Technology (IFPB-Sousa).

The samples collected were analysed for: pH, EC and K contraction$^+$, Na$^+$, Ca^{+2} , Mg^{+2} , SO$_4^2$ ',CO$_3^2$ ',HCO$_3$ ',Cr, CaCO$_3$, RAS, turbidity, temperature and dissolved oxygen (DO). RAS, RASaj and RAScor were also calculated according to Medeiros & Gheyi (1997) and Ayers & Westcot (1999).

The samples were always collected in the morning at the same reference point (water source). 100 mL of the sample was always collected in a previously prepared bottle for on-site analysis and another 2-litre bottle for laboratory analysis.

Samples for analysis of turbidity parameters were collected in polyethylene bottles, previously washed (neutral detergent, tap water and distilled/deionised water) and dried. Temperature, pH, dissolved oxygen (DO) and conductivity were measured in the field at the time of collection. Air and water temperatures were measured with thermometers at the time of each collection.

CHAPTER 4

RESULTS AND DISCUSSION

4.1. Characterisation of the river basin and water sources studied

Of the two communities studied, Várzea Comprida dos Oliveiras and Bezerro, there were a total of 18 (eighteen) tube wells, in which the first had 11 (eleven) wells and the other had 07 (seven), with one well used only for human consumption, 01 (one) Piranhas River analysed along with the others, the main body of water located between the communities.

Figure 10: Pump outlet with cover **Source:** author, 2014
Figure 11: Pump outlet without cover **Source:** author, 2014

According to data obtained from AESA - the Executive Agency for Water Management in the state of Paraíba, the main river in the federally-owned basin is the Piranhas-Açu river, which rises in the municipality of Bonito de Santa Fé, in the state of Paraíba and follows its natural course through the state of Rio Grande do Norte, until it flows into the Atlantic Ocean. All the wells are located in this important basin for the states of Rio Grande do Norte and Paraíba.

Figure 12: Rio do Peixe during the dry season **Source:** author, 2014
Figure 13: Piranhas River during the rain **Source:** author, 2014

The water collection points were geo-referenced using a GPS device and transferred to a map with geographical location data, which allowed the points to be obtained with latitude and longitude in UTM (Universal Transverse Mercator) projection.

With regard to the use of water for irrigation, the utilisation of this resource differs in terms of the flow (Q) of the water coming from the underground spring by means of pumps.

Table 2: Geographical location of the wells in the Bezerro Community / Pombal-PB

WELL	WELL NAME	GEOREFERENCING THE WELL	
1	A - Horta	S 06° 45.154'	W 037° 52.081'
2	A - Consumption	S 06° 45.083'	W 037° 52.075'
3	B	S 06° 44.990'	W 037° 52.168'
4	C	S 06° 45.982	W 037° 47.981
5	D	S 06° 45.144'	W 037° 52.318'
6	E	S 06° 45.032'	W 037° 52.408'
7	F	S 06° 45.027'	W 037° 52.738'

Source: Author, 2014

Table 3: Geographical location of the wells in the Várzea Comprida dos Oliveiras community / Pombal-PB

PORTION	WELL NAME	GEOREFERENCING THE WELL	
1	A	S 06° 44.842'	W 037° 51.709'
2	B	S 06° 44.814'	W 037° 51.761'
3	C	S 06° 44.876'	W 037° 51.640'
4	D	S 06° 44.935'	W 037° 51.660'
5	E	S 06° 45.199'	W 037° 51.568'
6	F	S 06° 45.192'	W 037° 51.533'
7	G	S 06° 45.471'	W 037° 51.656'
8	H	S 06° 45.445'	W 037° 51.832'
9	I	S 06° 45.277'	W 037° 51.852'
10	J	S 06° 45.294'	W 037° 51.147'
11	L	S 06° 45.475'	W 037° 51.458'

Source: author, 2014.

Through direct contact with the producers, it was possible to see that problems such as water scarcity are the most serious, especially during periods of drought, jeopardising agricultural activities due to the low level of water in the river and wells.

During the visits and water sampling, the presence of fertiliser and pest control packaging and rubbish discarded in the vicinity and on the fields was noted, as was a survey of the owners about the use of these chemicals on the crops during planting.

Figure 14: CYPERPOUR 15 **Source:** author, 2014
Figure 15: Rubbish dumped in the open **Source:** author, 2014

4.1.1 Crops grown on the properties

Through direct contact with producers, it was possible to identify the main crops produced in the communities. The crops' total need for water depends essentially on climatic conditions, the length of the cycle and the cultivation and irrigation systems adopted. Irrigation efficiency depends mainly on the characteristics of the system, the type of soil and water management.

Culturas

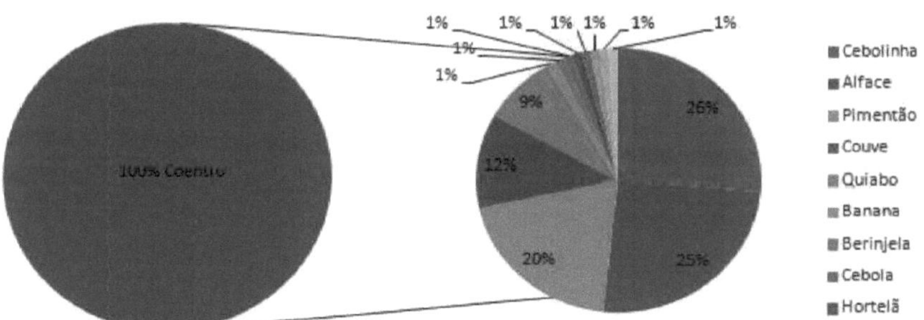

Figure 16: Cultivated crops

Source: author, 2014

4.1.2 Water Sample Collection

The water samples were collected from natural springs and tube wells during the dry season between 11/2013 and 05/2014.

In the communities under study, farmers and the President of the Rural Producers' Association, or another person with knowledge of the area, were contacted to help recognise the site. Samples were collected using a plastic pet bottle with a diameter of 10 mm and a length of 22 mm (Figure 17).

Figure 17: Bottle used to collect water
Source: author, 2014.

The water was stored in plastic bottles, previously washed and rinsed at the time of collection, which were completely filled, sealed and labelled with an identification number, the name of the community and the type of source to which they belonged (Figures 18 and 19).

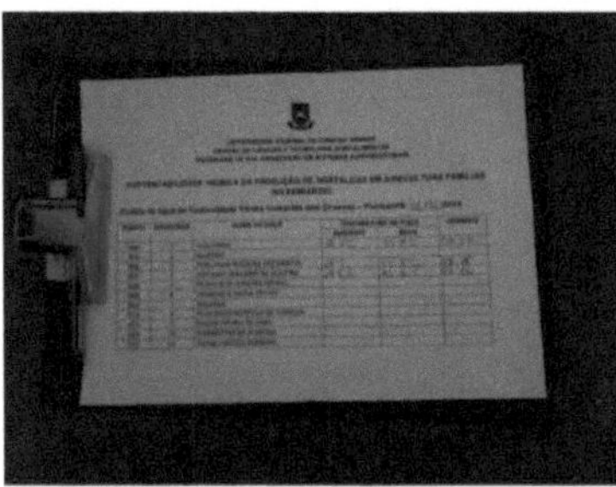

Figure 18: Field note form **Source:** author, 2013.
Figure 19: Water samples **Source:** author, 2013.

The results were compared mainly with the values established by CONAMA (National Environment Council) Resolution No. 357 of 17 March 2005 for Class 1 water, which is intended for irrigating vegetables consumed raw.

Class 1:
-to the   supply for human consumption, after simplified treatment;
-the   protection of aquatic communities;
   -primary contact recreation, such as swimming, water skiing and diving;
   -irrigation of vegetables that are eaten raw and fruit that grows close to the ground and is eaten raw without removing the skin;
-the   protection of aquatic communities on indigenous lands.

Table 4 - Values for Class 1 water according to CONAMA Resolution 375/05

Parameters	Values
PH	6,0 a 9,0
OD	Not less than 6 mg L^{-1}
Turbidity	Up to 40 nephelometric units (NMP)

4.1.3. Flow (Q)

The areas studied are located in a hydrographic basin that influences the flow of the wells studied, which varies the amount of water present according to the season.

To calculate the average flow rate of the wells, we used a graduated bucket with an identified volume of 20 litres, and a digital stopwatch to measure the average time taken to fill the bucket in triplicate, then using the following equation we can calculate the possible flow rate (Q) of the wells.

Q = V/tç

Where: Q= flow;

V= volume and t = time

Figure 20: Volumetric bucket **Source:** author, 2014.
Figure 21: Volume measurement **Source:** author, 2014.

Table 5: Flow (Q) of the wells in the Bezerro Community / Pombal-PB

WELL	WELL NAME	Pump flow Q = V/t	
		Qaverage L/ S	Qaverage m /S
1	A- Horta	0,3896	0,0003896
2	A - Consumption	0,3109	0,0003109
3	B	0,5882	0,0005882
4	C	0,8333	0,0008333
5	D	0,7228	0,0007228
6	E	0,2817	0,0002817
7	F	2,1436	0,0021436

Source: author, 2014

Table 6: Flow (Q) of the wells in the community of Várzea Comprida dos Oliveiras/Pombal-PB

WELL	WELL NAME	Pump flow Q = V/t	
		Average Q L/s	Average Q m /S

1	A	1,0086	0,0010086
2	B	0,8449	0,0008449
3	C	0,4698	0,0004698
4	D	1,1940	0,001194
5	E	0,2880	0,000288
6	F	1,1376	0,0011376
7	G	1,2285	0,0012285
8	H	1,6835	0,0016835
9	I	1,3898	0,0013898
10	J	0,7719	0,0007719
11	L	3,1596	0,0031596

Source: author, 2014.

4.1.4 pH values of well water

The pH values of the water samples from the wells were compared with the parameters established by CONAMA-MMA resolution no. 357 of 17/03/2005, which provides for the classification of bodies of water and environmental guidelines for their classification, which is indicated from 6 to 9. For most natural waters, pH is influenced by the concentration of H+ originating from the dissociation of carbonic acid, which generates low pH values, and the reactions of carbonate and bicarbonate ions with the water molecule, which raise pH values to the alkaline range (ESTEVES, 1998).pH indicates the intensity of acidity or alkalinity. In surface waters (rivers and lakes) this value is influenced by various factors such as the geology of the region where the body of water is located and possible sources of pollution (dumping of domestic, industrial or agricultural effluents) (FRAVET, 2006).

4.1.5 Air and water temperature.

Temperature plays a fundamental and important role in controlling the aquatic environment, conditioning the influences of a series of physical parameters such as viscosity, surface tension, compressibility, specific heat, ionisation constant, latent heat of vaporisation, thermal conductivity and vapour pressure. Surface temperature is influenced by factors such as latitude, altitude, season, time of day, flow rate and depth.

Temperature variations in air and water are important factors in the energetic and ecological reactions applied to water resources, exerting a direct influence on the various types of organisms and on the content of dissolved gases in the water (BRANCO, 1986).

The following table shows the air and water temperature values recorded during the water collection period in the two communities studied.

4.1.6 Dissolved Oxygen (DO).

Dissolved oxygen (DO) is an important chemical variable for environmental conditions. Although it is not a parameter used to characterise water quality for irrigation, measuring dissolved oxygen concentration detects the effects of oxidisable waste on receiving waters and the efficiency of sewage treatment during biochemical oxidation (CETESB, 2009).

The determination of (DO) provides information on the biochemical and biological reactions taking place in the water, as well as indicating the capacity of the body of water to promote self-depuration. The concentration of dissolved oxygen varies according to temperature, altitude and aeration.

Although in practice it is not a parameter used to characterise water quality for irrigation, it can be indicative of pollution, the concentration of dissolved solids and organic matter in the water (LARCHER, 2000; MORAES, 2001; VON SPERLING, 2005).

CONAMA Resolution 357/05 stipulates that in any sample collected, the dissolved oxygen values for class 1, 2 and 3 water cannot be less than 6, 5 and 4 mg/L, respectively.

Table 7 - Variation of dissolved O2 in relation to temperature

Relationship between temperature and dissolved oxygen (DO) solubility

Temperature (°C)	Oxygen solubility (mg / L)
0	14,6
5	12,8
10	11,3
15	10,2
20	9,2
25	8,6
100 boiling	0

Class 1
- for human consumption after simplified treatment;
- the protection of aquatic communities;
- primary contact recreation, such as swimming, water skiing and diving;
- *for irrigating vegetables that are eaten raw and fruit that grows close to the ground and is eaten raw without removing the skin;*
- the protection of aquatic communities on indigenous lands.

Class 2
-to the supply for human consumption, after conventional treatment;
-the   protection of aquatic communities;
   -primary contact recreation, such as swimming, water skiing and diving;
   -*irrigation of vegetables, fruit plants and parks, gardens, sports and leisure grounds, with which the public may come into direct contact',* and
   -aquaculture and fishing.

Class 3
-to the supply for human consumption, after conventional or advanced treatment;
   -*irrigation of tree, cereal and fodder crops;*
   -amateur fishing;
   -secondary contact recreation;
   -animal watering.

4.1.7 Turbidity.

The turbidity of a water sample is the degree of attenuation in intensity that a beam of light undergoes when it passes through it (this reduction is due to absorption and scattering, since the particles that cause turbidity in water are larger than the wavelength of white light), due to the presence of suspended solids, such as inorganic particles (sand, silt, clay) and organic debris, such as algae and bacteria and plankton in general (CETESB, 2009).

Turbidity was read directly on the SL 2K turbidity meter, following the instructions in the device's manual, expressed in Nephelometric Turbidity Units (UNT).

Turbidity is one of the parameters used to assess the quality of water for human consumption, but it doesn't have much influence when it comes to using water for irrigation.

According to Carvalho, 1994. This parameter can be used to measure the concentration of suspended sediment, which is of great importance for the quality of irrigation water.

4.1.8. Cor.

Table 8 - Colour measurement of the wells in the Bezerro Community - Pombal/PB

WELL	NAME	MEASURED VALUE g/L
1	A - Horta	0,04
2	A - Consumption	0,06
3	B	0,02
4	C	0,02
5	D	0,00
6	E	0,04
7	F	0,03
8	PIRANHAS RIVER	0,24

Source: author, 2014

Header 9 - Colour measurement of the wells in the Várzea Comprida dos Oliveiras Community - 'ombal/PB

WELL	NAME	MEASURED VALUE g/L
1	A	0,006
2	B	0,02
3	C	0,00
4	D	0,00
5	E	0,01
6	F	0,00
7	G	0,00
8	H	0,02
9	I	0,006
10	J	0,02
11	L	0,02

Source: author, 2014

4.1.9 Descriptive data analysis

There was an increase in the pH of the water from the wells in the Bezerro community (Figure 22) between 27/12/2013 and 07/02/2014, with wells B and A having the highest values (8.16 and 7.8) respectively (Figure 22). Water with high pH values can harm irrigation systems, as salts at this pH end up precipitating in the water, causing blockages in the pipes and clogging nozzles and emitters. Wells C, D, E,
F and G also showed the same behaviour as wells A and B (Figure 22), with values increasing up to the respective date (07/02/2014).

The Piranhas River, on the other hand, showed little variation in water pH during the monitoring period, with pH ranging from 6.98 to 6.5 between 27/12/2013 and 17/04/2014, with a tendency to drop sharply from 27/03/2014 onwards (Figure 22). These values are very close to those observed by Femandes and Santiago (2001), which were 6.7 and 7.13. Although most waters have a pH just above 7.0, knowing that it was measured at the pump outlet when the irrigation system was running, when the system is stopped, it tends to increase, causing the ISL to increase, consequently the solubility of carbonates decreases and precipitation and subsequent clogging of emitters can occur more easily, a fact that confirms what was written by Egrega Filho et al. (1999) and Medeiros (2003).

The water from the wells in the community of várzea comprida dos Oliveiras showed similar trends as shown in figure 23, with a small increase followed by a decrease in the results between 26/12/2013 and 06/02/2014. The pH values varied from 5.23 in well D on 06/02/2014 to 7.73 in well A on 16/01/2014. Medeiros et al. (2003) evaluated the pH in ten wells between the municipalities of Mossoró-RN and Baraúna-RN, and observed a pH below 7.

Lima et al. (2007), working with castor beans, used water from a well near Mossoró-RN, which had a pH of 7.3 (pH 1:2.5). Rocha (2008), studying the quality of tubular well water in the Rio do peixe basin, found pH values ranging from 6.85 to 8.27. In a study carried out by Alencar (2007) in the agricultural centre of Mossoró-RN (Pau Branco, Califórnia, Gangorra, Posto Fiscal) and Baraúna (Poço Baraúna, Catiguinha and Olho D'água da Escada), he found results in the 6.33 to 7.36 range. Similar results were seen by Oliveira and Maia (1998), with pH values of 6.9 to 7.6, with the well in the city of Governador showing 8.2.

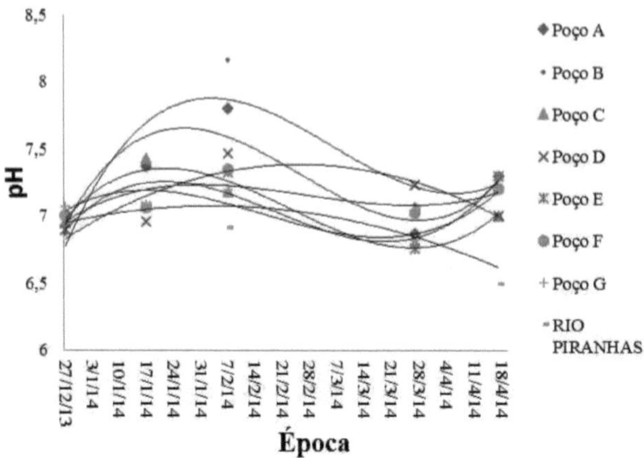

Figura 22   - pH of the water from the wells in the Bezerro community and the Piranhas river

Treatments	Equations	R^2
Well A	$6E\text{-}06x^J - 0.7956x^2 + 33174x - 5E\text{+}08$	0,7892
Well B	$6E\text{-}06x^5 - 0.7357x^2 + 3068Ix - 4E\text{+}08$	0,7762
Well C	$5E\text{-}06x^J - 0.5815x^2 + 24243x - 3E\text{+}08$	0,9304
Well D	$-0.0002x^2 + 12.658x - 263877$	0,732
Well E	$5E\text{-}06x^J - 0.613 1x^2 + 25559x - 4E\text{+}08$	0,7069
Well F	$2E\text{-}06x^J - 0.2402x^2 + 10016x - 1E\text{+}08$	0,5397
Well G	$2E\text{-}06x^J - 0.2402x^2 + 10016x - 1E\text{+}08$	0,5397
Piranhas River	$-9E\text{-}05x^2 + 7.1654x - 149296$	0,6014

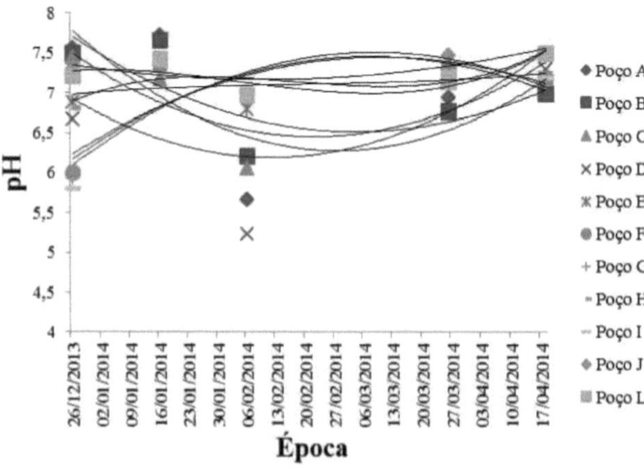

Figura 23   -pH of the water from the wells in the várzea comprida das Oliveiras community

Treatments	Equations	R^2
Well A	$0.0004x^2 - 30.823x + 642623$	0,4419
Well B	$0.0003x^2 - 21.524x + 448797$	0,5393
Well C	$0.0003x^2 - 25.75Ix + 536810$	0,5784

Well D	$0.0003x^2 - 26.427x + 550767$	0,3216
Well E	7,3	
Well F	$-0.0002x^2 + 18.415x - 384017$	0,6634
Well G	$-0.0003x^2 + 20.938x - 436616$	0,6474
Well H	$3E-06x^j - 0.393\,1x^2 + 16387x - 2E+08$	0,6067
Well I	$-0.0003x^2 + 22.371x - 466505$	0,633
Well J	$9E-05x^2 - 7.2567x + 151235$	0,4289
Well L	$2E-06x^j - 0.2892x^2 + 12052x - 2E+08$	0,6591

Figure 24 shows the monitoring of the water's electrical conductivity (EC) in the wells of the Bezerro community, with behaviour similar to that of pH, with an increase in EC until 07/02/2014 and a reduction from that date in most wells, showing that there is a relationship between the two variables.

The well showed the highest EC values, ranging from 0.66 to 0.92 dS m[1] , followed by wells G and D, which showed results of 0.57, 0.55 and 0.87 and 0.88 dS.m[1] . The Piranhas River showed the lowest EC results (0.08 to 0.22 dS.m[1]), with decreasing behaviour over time, obtaining the lowest value in the last collection (17/04/2014).

This trend coincided with the rainy season in the region, which runs from February to April. According to the information shown in figure 25, the electrical conductivity values of most of the wells in the várzea comprida da Oliveira community showed a reduction from 06/02/2014 onwards, so we can see the effect of the rains in reducing the electrical conductivity of the water, as has already been seen with the wells in the Bezerro community.

With the increase in rainfall there is a recharge in the amount of water in the wells, so the salts present in the water are diluted with the increase in the volume of water in the well. Well I showed a different behaviour, with an increase between 26/12/2013 and 16/01/2014 (0.67 to 0.85 dS m[1]) respectively, followed by a reduction until 06/02/2014 (0.73 dS m[1]), keeping the values constant, and a reduction from 26/03/2014 (0.6 dS m').[1]

The water from the wells is classified as having moderate salinity (0.7 to 3.0 dS m[1]) (AYERS; WESTCOT, 1985).

The intense evapotranspiration in the Brazilian semi-arid region, which is often greater than the annual rainfall, has compromised the quality of reservoir water (SILVA et al., 2004), increasing the concentration of salts in surface water.

Medeiros et al. (2003) observed ECs of 1.17 to 2.98 dS m^{-1} in different wells in the Mossoró - Barauna-RN region. Similarly, Oliveira and Maia (1998), observing the electrical conductivity of wells in a town in the west of Rio Grande do Sul, found values ranging from 0.9 dS m^{-1} in Ipanguaçu to 4.0 dS m^{-1} in Grossos, a result that may have been influenced by the proximity of the wells to the sea. Andrade Júnior et al. (2006) assessed the water quality of 225 wells in the semi-arid region of the state of Piauí during the dry season and found that the highest EC values were found in the eastern part of the area studiedSantiago et al. (1999), in a study carried out in Picos, found values of 2.5 dS m^{-1} at 25°C, showing that these waters are restricted for agricultural activities due to the risk of soil salinisation.

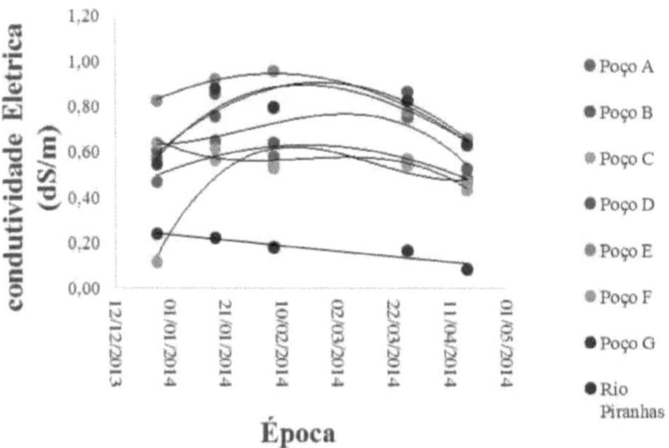

Figura 24 - Electrical conductivity of water in wells in the community of Bezerro and the Piranhas River.

Treatments	Equations	R^2
Well A	$-5E-05x^2 + 3.805Ix - 79314$	0,6628
Well B	$-8E-07x^J + 0.1017x^2 - 4238,Ix + 6E+07$	0,5567
Well C	$-8E-07x^J + 0.1057x^2 - 4408x + 6E+07$	0,8665
Well D	$-9E-05x^2 + 7.5827x - 158074$	0,7595
Well E	$-6E-05x^2 + 5.3133x - 110721$	0,9845
Well F	$1E-O6x^j - 0.1817x^2 + 7580.5x - 1E+08$	0,9053
Well G	$4E-07x^J - 0.0494x^2 + 2065.2x - 3E+07$	0,7611
Piranhas River	$-0,0012x + 50,324$	0,868

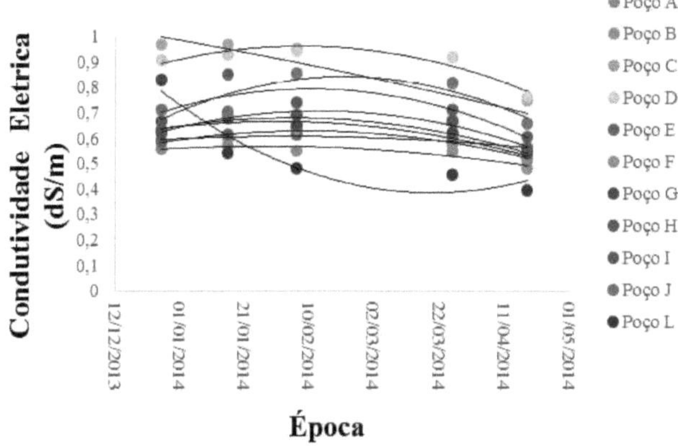

Figura 25 - **Electrical conductivity of the water in the wells of the community of várzea comprida de Oliveiras.**

Treatments	Equations	R^2
Well A	$-1E\text{-}O5x^2 + 0.9987x - 20805$	0,823
Well B	$-8E\text{-}06x^2 + 0.6637x - 13829$	0,9151
Well C	$-4E\text{-}06x^2 + 0.2917x - 6025.1$	0,8052
Well D	$-4E\text{-}05x^2 + 3.0374x - 63295$	0,8825
Well E	$-2E\text{-}05x^2 + 1,9062x - 39724$	0,891
Well F	$-2E\text{-}05x^2 + 1.8803x - 39176$	0,8813
Well G	$-3E\text{-}05x^2 + 2.1981x - 45803$	0,8462
Well H	$-3E\text{-}05x^2 + 2.8396x - 59181$	0,9418
Well I	$-4E\text{-}05x^2 + 3.5862x - 74736$	0,6615
Well J	$5E\text{-}05x^2 + 4.2734x - 89081$	0,6701
Well L	$6E\text{-}05x^2 - 4.7914x + 99944$	0,8897

The waters of the Piranhas River showed a different behaviour to the other well waters (Figure 26). There was a reduction in water turbidity between 27/12/2013 and 07/02/2014, followed by an increase in values until 17/04/2014. The highest turbidity values (44.8 and 166 mg Pt / L) found in the water of the Piranhas River coincided with the rainy season. This can be explained by the fact that the water coming from the rain would cause a movement of solid particles from the river bottom, as well as the dragging of coarse matter from the river bed. Among the wells in the Bezerro community, well B showed the highest results, with values of 7.34 and 6.33 mg Pt / L between 07/02/2014 and 17/04/2014 respectively.

For the wells in the community of várzea comprida da Oliveiras, the highest values were found in the collection of 16/01/2014 (figure 27), with well J showing 2.55 mg Pt / L. Wells F and E showed the lowest values on 26/03/2014 (0.18 to 0.23 mg Pt / L). Similar results were observed by Barcellos et al. (2000) and Silva, Araújo and Souza (2007) studying the turbidity of shallow wells in the municipality of Lavras-MG, where they found values of 2.52 mg Pt / L and 5 mg Pt / L.

Normally, groundwater is free of colour, but atypically it can reach values of up to 100 mg Pt/L (FEITOSA E FILHO, 1997), which causes psychological repulsion for the consumer. Rocha (2008), assessing the quality of well water in the state of Piauí, observed that turbidity had a maximum value of 787 mg Pt/L. Figure 26 - Turbidity of the water from the Bezerro and Piranhas river wells.

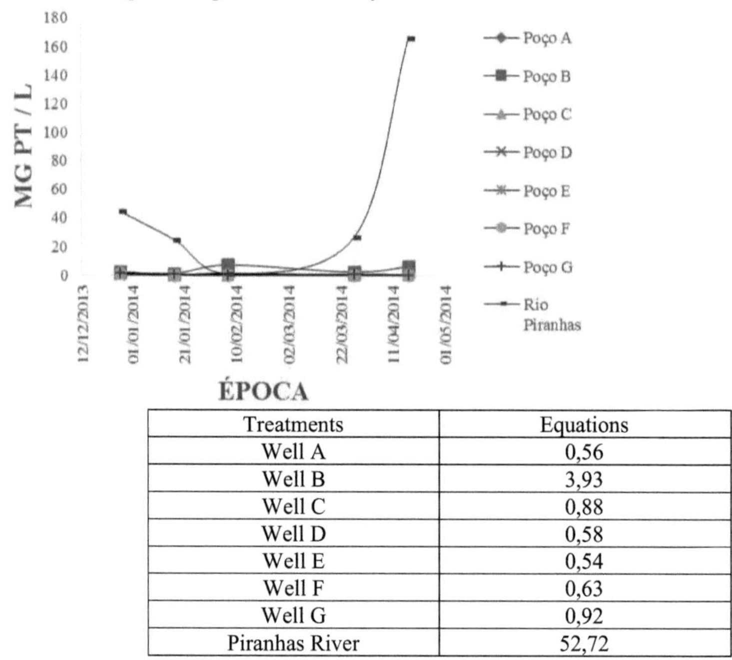

Treatments	Equations
Well A	0,56
Well B	3,93
Well C	0,88
Well D	0,58
Well E	0,54
Well F	0,63
Well G	0,92
Piranhas River	52,72

Figura 27: - Turbidity of the water in the wells of the Oliveiras long floodplain

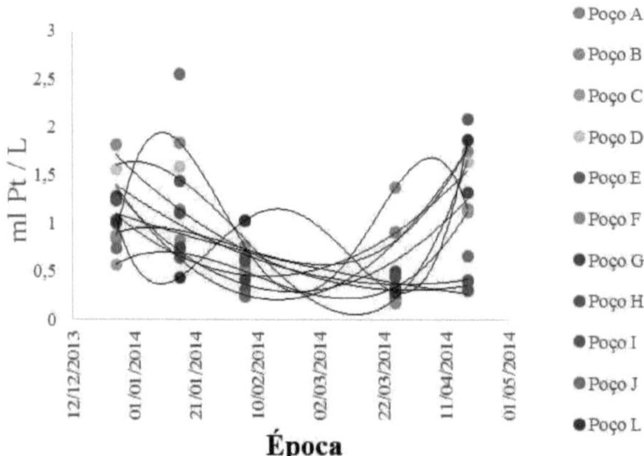

Treatments	Equations	R^2
Well A	$4E\text{-}06x^J - 0.5542x^2 + 23097x - 3E\text{+}08$	0,8772
Well B	1,56	
Well C	$5E\text{-}06x^j - 0.5979x^2 + 24922x - 3E\text{+}08$	0,9975
Well D	$9E\text{-}06x^3 - 1,1008x^2 + 45877x - 6E\text{+}08$	0,9756
Well E	$0.0004x^2 - 35.605x + 742128$	0,706
Well F	$-1E\text{-}O7x^j + 0.0178x^2 - 750.52x + 1E\text{+}07$	0,862
Well G	$0.0003x^2 - 26.llx + 544275$	0,9644
Well H	$= 3E\text{-}05x^2 - 2.2702x + 47476$	0,8762
Well I	$0.0001x^2 - 10.08x + 210292$	0,6646
Well J	$0.0003x^2 - 22.221x + 463120$	0,4526
Well L	1,6	

It was noted that the waters of the wells in the Bezerro community showed similar behaviour between the collection times for dissolved oxygen, with a small increase in values between 27/12/2013 and 17/01/2014 (figure 28), followed by a sharp increase between 17/01/2014 and 07/02/2014. Well A and the Piranhas River showed the highest values (22.4, 21.5 and 22.2, 20, 7) between 07/02/2014 and 27/03/2014. On the other hand, the wells in the várzea comprida da Oliveiras community showed a large increase between 16/01/2014 and 06/02/2014, with values ranging from 20.6 to 21.1 mg/L, keeping the values almost constant (figure 29). Well H showed an oscillation between values, with a reduction between 06/02/2014 and 26/03/2014 (20.7 and 15.5 mg/L), followed by an increase to 20.2 mg/L (28/04/2014).

A study carried out by Barcellos et al. (2000) on spring waters in the municipality of Lavras, Minas Gerais, found values of 5.2 mg/L for shallow wells. Similar results were found by Manchester et al., (2013) evaluating mine waters in the municipality of Teofilo Otoni-MG, who found results close to 6 mg / L. Nascimento and Barbosa (2005) found values of 6.4 mg / L in the waters of the Caia basin in the municipality of Salvador-BA.

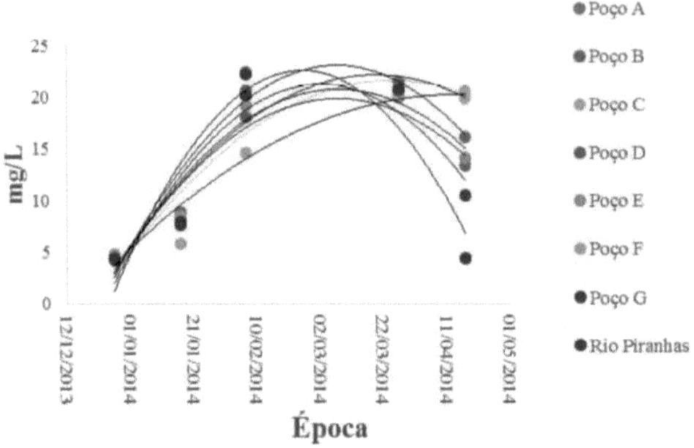

Treatments	Equations	R^2
Well A	$-0.0041x^2 + 343.88x - 7E+06$	0,8922
Well B	$-0.0027x^2 + 228.87x - 5E+06$	0,9061
Well C	$-0.0024x^2 + 201.67x - 4E+06$	0,8473
Well D	$0.0034x^2 + 281.3x - 6E+06$	0,9083
Well E	$-0.0036x^2 + 296.22x - 6E+06$	0,896
Well F	$-0.0014x^2 + 118.79x - 2E+06$	0,998
Well G	$- 0.0045x^2 + 374.3x - 8E+06$	0,8427
Piranhas River	$-0.006x^2 + 499.66x - 1E+07$	0,7859

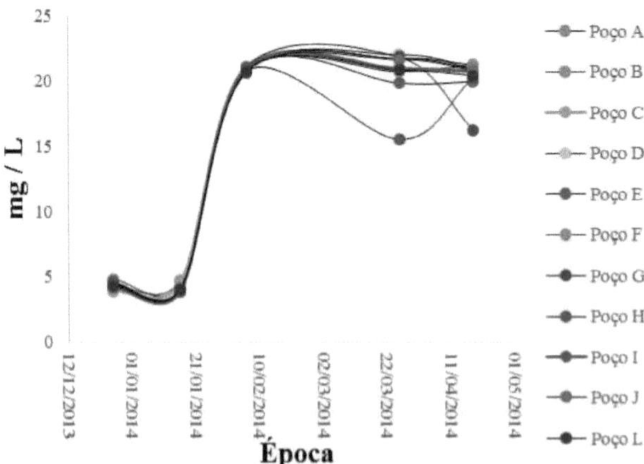

Treatments	Equations
Well A	14,56
Well B	14,34
Well C	14,24
Well D	14,32
Well E	13,84
Well F	14,46
Well G	13,44
Well H	12,98
Well I	14,12
Well J	14,3
Well L	14,04

The water temperature of the wells in the Bezerro community remained constant during the monitoring period, with a slight reduction in well C (41.2 to 33.9 °C) between 27/12/2013 and 17/01/2014 and 27/03/2014 and 17/04/2014, while well G showed an increase between 27/03/2014 and 17/04/2014, showing values of (30 to 33.7 °C) (figure 30). The highest values were found at the start of monitoring, with a value of 41 °C in well C. The wells in the várzea comprida da Oliveiras community showed the lowest values between 26/12/2013 and 16/01/2014 and from 26/03/2014 onwards (figure 31). Well D was the treatment that obtained the highest results, with an average of 35 °C. Wells E and F showed the lowest average values of degrees Celsus among the wells (3.54 and 3.56 °C).

In a study carried out by Barcellos et al. (2000) on the waters of shallow wells in Lavras-MG, they found average values of 23 °C. Similar results were found by Nascimento and Barbosa (2005), who assessed the water quality of a groundwater aquifer in the city of Salvador and found an average value of 26.5 °C.

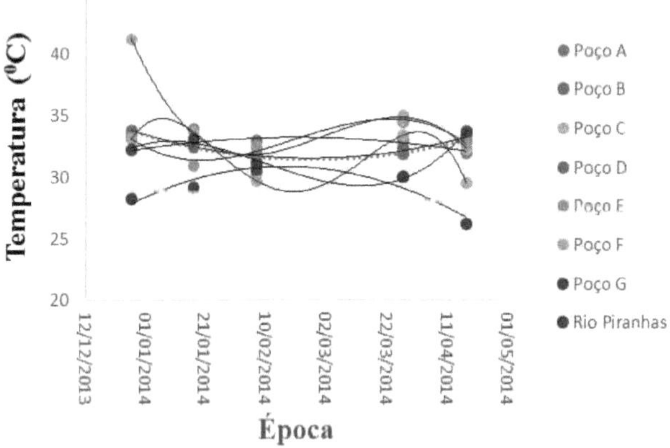

Figura 30: water temperature of the wells in the calf community and the Piranhas River

Treatments	Equations	R^2
Well A	$-0.0004x^2 + 29.402x - 612856$	0,9009
Well B	$0.0007x^2 - 54.385x + 1E+06$	0.9649
Well C	$-5E-05x^J + 6.5749x^2 - 274168x + 4E+09$	0,9955
Well D	$0.0006x^2 - 54.194x + 1E+06$	0,4202
Well E	$-3E-05x^J + 3.3963x^2 - 141589x + 2E+09$	0,9392
Well F	34,6	
Well G	$3E-O5x^5 - 3.6252x^2 + 151096x - 2E+09$	0,9784
Piranhas River	$-0.0012x^2 + 95.987x - 2E+06$	0,8593

Figura 31: water temperature of the wells in the várzea comprida da Oliveiras community

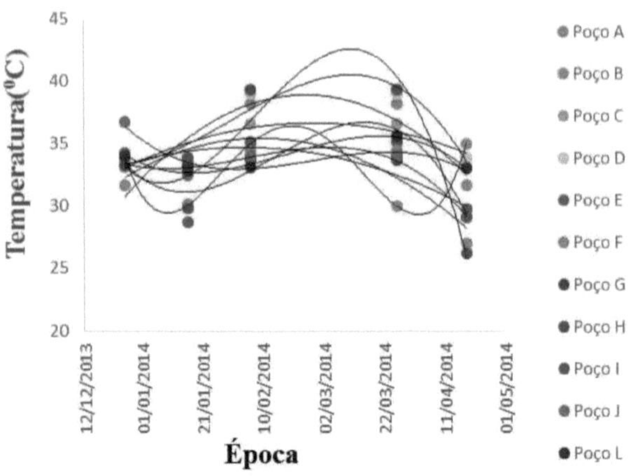

Treatments	Equations	R^2
Well A	$-0.0023x^2 + 188.36x - 4E+06$	0,865
Well B	32,4	
Well C	$-0.0009x^2 + 77.748x - 2E+06$	0,7102
Well D	$-3E-05x^3 + 3.8234x^2 - 159320x + 2E+09$	0,8815
Well E	$-5E-05x^J + 6.6609x^2 - 277649x + 4E+09$	0,6725
Well F	$= -0.0016x^2 + 129.63x - 3E+06$	0,7908
Well G	$-2E-05x^J + 2.6978x^2 - 112430x + 2E+09$	0,9931
Well H	$= -9E-05x^J + 10.875x^2 - 453224x + 6E+09$	0,8885
Well I	$= -0.0009x^2 + 73.41 Ix - 2E+06$	0,7711
Well J	$2E-05x^j + 2.7577x^2 - 114984x + 2E+09$	0,8018
Well L	$-3E-05x^J + 3.1305x^2 - 130501x + 2E+09$	0,9724

31.1.10. Descriptive analyses of chemical analysis data

According to the results obtained, bicarbonate (HCO3') was the anion that showed the highest results in all the wells and in the Piranhas River, with values ranging from 11.32 $mmol_c$ / L in well 4 (figure 32D) to 1.58 $mmol_c$ L^{-1} (figure 32T). Wells 18, 10 and 6 had HCO3 concentrations of 11, 10.42 and 10.34 $mmol_c$ / L. Medeiros el at (2003) found HCO3 values above 5 $mmol_c$ L^{-1} when studying salt concentrations in wells in the Mossoró-Baraúna region.

Results above 5 $mmol_c$ L^{-1} in irrigation water can cause phosphate fertilisers to precipitate when added, causing emitters to clog. In water containing high concentrations of HCO_3 ions ", there is a tendency for calcium and magnesium to precipitate in the form of carbonates, thus reducing the concentrations of these ions and increasing the proportion of sodium (BERNARDO, SOARES, MANTOVANNI, 2011). Similar results were found by Alencar (2007), who assessed the quality of water from wells in the Jandaira limestone and found an average value of 4.0 $mmol_c$ / L.

Silva et al. (2003) assessed the chemical quality of well water in the south-eastern region of the state of Piauí and found values of 134, 170 and 313 mg/L of HCCh for wells in the municipalities of Padre Marcos, Picos and Pio IX respectively. Lima et al, (2007) evaluated the effect of salinity on castor bean cultivation and obtained six levels of salinity in the irrigation water (0.5, 1.0, 1.5, 2.0 and 2.5 dS m^{-1}), and concentrations of HCO3' (2.3, 4.0, 3.3, 3.5, 3.3, 3.7 $mmol_c$ L^{-1}), thus observing a relationship between the two variables, a small increase in the HCO3' content with increasing salinity.

Carbonate (CO_3^{2-}) showed the highest values in wells 11, 18 and 10 (Figure 32L, 32S and 32J), with 1.34, 1.14 and 1.12 $mmol_c$ L^{-1} , and the lowest was found in the water of the Piranhas River, with 0.0 mmolc / L. Wells 2, 5 and 17 showed results of 0.72, 0.64 and 0.94 mmol L_c^{-1} .Rocha (2008) assessed the concentration of CO_3^{2-} in well water in the Rio dos Peixes basin and found 0.0 $mmol_c$ L^{-1} . However, the pH value measured in the field was 8.41 in the town of Canabrava, indicating the presence of carbonates, which did not occur in the laboratory analysis. Their absence can be explained by the reaction between the carbon dioxide present in the atmospheric air and the carbonates to form bicarbonates.

Among the ions studied in this study, potassium and sulphate showed the lowest results among the wells, with the exception of the Piranhas River, where carbonate showed the lowest value. Potassium values ranged from 0.02 to 0.12 mmolc / L in wells 12 and 16 (Figure 32M and 32Q). Potassium showed the lowest values among the ions analysed, with an average value of 0.07 mmol / L_c

Alencar (2007), monitoring the quality of well water in the Mossoró-Baraúna region, found

0.06 mmol$_c$ / L of potassium in wells in the municipality of Baraúna and 0.41 mmolc / L in wells in the Gangorra community in the municipality of Mossoró. Costa and Gheyi (1984) observed potassium values of 0.03 to 0.31 in wells in the municipalities of São Bento-PB and Brejo do Cruz-PB. Lima et al. (2007) observed a relationship between EC

(electrical conductivity) of the water with the potassium content, obtaining five levels of saline water (0.5, 1.0, 1.5, 2.0, 2.5 dS m^{-1}) in the work, the potassium values increased linearly (0.24, 0.34, 0.40, 0.46, 0.51 and 0.58 mmol$_c$ / L).

Sulphate showed results ranging from 0.05 mmol$_c$ / L in well 5 (Figure 32E) to 0.86 mmol$_c$ / L in well 3 (Figure 32C). Sulphate showed an average value of 0.21 mmol$_c$ / L. The presence of one ion in excess can cause a deficiency or inhibit the absorption of another due to precipitation.

Excess sulphate can cause calcium precipitation, affecting plant growth due to a lack of the precipitated element rather than an excess of another ion. According to Pizarro (1978), banana and lettuce crops are affected by high sulphate levels in irrigation water. In a study carried out by Costa and Gheyi (1984) assessing the quality of water in wells in the micro-region of the municipality of Catolé do Rocha-PB, they observed sulphate values of 0.5 mmol$_c$ / L in wells in Jerico-PB and 1.00 mmol$_c$ / L in Bom sucesso.

The ions calcium (Ca^{2+}) and magnesium (Mg^{2+}) were on average the second and third most concentrated in the waters evaluated in this study. Calcium showed values of 3.3 mmol$_c$ / L in well 16 (Figure 32Q), 2.4 mmol$_c$ / L in wells 5 and 6 (Figures 32E and 32F), 1.4 mmol$_c$ / L in well 10 (Figure 32J) and 0.9 mmol$_c$ / L in well 11 (Figure 32L), with an average value of 1.9 mmol$_c$ / L. The Mg ion^{2+} also showed the highest and lowest results in wells 16 and 11 (Figure 32Q and 32L), as was seen for Ca^{2+} , thus showing a relationship between these two elements in the water. The levels of Ca^{2+} and Mg^{2+} in the irrigation water are affected by the concentration of bicarbonate in the water, since an increase in the bicarbonate content in the irrigation water will lead to the formation of precipitates, in this case calcium and magnesium carbonate (CaCCh and MgCCh).

Medeiros et al., (2003) found that in all the wells analysed the calcium concentration was quite high, with the waters of the Boa Agua, Cacimba Funda, Mata Fresca and Serra Mossoró communities showing the highest values for these ions. Medeiros et al. (2003) also showed a table with a relationship between water salinity and Ca+Mg content, obtaining an equation Y = 4.75X + 4.56, Martins (1993) and Silva Júnior et al. (1999) found similar results with other waters in the north-east of Brazil. Oliveira and Maia obtained similar results to this work, studying the concentration of salts in wells in the Apodi plateau, finding an average value of 3.33 mmol$_c$ / L of calcium and 2.9 mmol$_c$ / L of magnesium.

Júnior, Gheyi and Medeiros (1999) observed Ca values (1.47, 4.2 and 2.75 mmol$_c$ / L) in wells in the municipalities of Pau dos Ferros-RN, Boqueirão-PB and Picuí-PB, respectively, and 1.08, 4.96

and 5.70 mmol$_c$ / L of Mg^{2+} in the respective municipalities. [a]Alencar (2007) observed Calcium values that varied between 6.2 and 26.5 mmolc.L^{-1} in the 1st collection and 7.1 and 23.7 mmolc.L^{-1} in the 4th collection[a] , where only the town of Gangorra showed values considered to be above normal (>20 mmolc.L^{-1}). With regard to Mg , the values varied between 2.69 and 8.53 mmolc.L^{-1} in the I[a] reading and between 2.6 and 15.4 mmolc.L^{-1} in the 4[a] , with the Gangorra site showing the highest values in both seasons. These values are considered by Ayers & Westcot, (1999) to be above normal (0-5 mmolc.L)$^{-1}$

Chloride (Cl$^-$) and sodium (Na$^+$) in the water from the wells and the Piranhas River showed average values of 1.5 and 1.67 mmol$_c$ L^{-1} simultaneously. Well 13 showed the highest value of Cl$^-$ (2.5 mmol$_c$ L^{-1}), while well 1 showed the lowest value of Cl$^-$ (0.6 mmol$_c$ L)$^{-1}$

Similar results were found by Júnior, Gheyi and Medeiros (1999), Medeiros et al., (2003), where they found Cl values$^-$ varying between 4.4 and 6.9 mmol$_c$ L^{-1} .

Chloride is neither retained nor adsorbed by the soil particles and moves easily with the water in the soil, but is absorbed by the roots and translocated to the leaves, where it accumulates through transpiration. The sensitivity of crops to this ion varies greatly, such as fruit trees, which begin to show symptoms of damage at concentrations above 0.3% chloride on a dry weight basis, while tolerant species can accumulate up to 4.0% to 5.0% chloride without showing any symptoms of toxicity (DIAS, BLANCO, 2010).

Alencar (2007), studying salt concentrations in irrigation water, observed that almost all the wells in the Mossoró-Baraúna region showed chloride levels above 10 mmol$_c$ L^{-1} , and he also observed a relationship between EC x Cl$^-$, because with the linear increase in Cl' concentration there was also a linear rise in the electrical conductivity curve. Maia, Morais and Oliveira (1998) also observed a strong relationship between EC x Cl$^-$, with R^2 equal to 0.85.

Sodium obtained the highest results in wells 3, 4 and 15, with values of 3.68, 3.08 and 2.97 mmolc L^{-1} , and the lowest was found in well 14. Similar results were noted by Medeiros et al. (2003), where values ranged from 2.9 to 4.9 mmol$_c$ L^{-1} .

Sodium toxicity is more difficult to diagnose than chloride, but it has been clearly identified as a result of a high proportion of sodium in the water. Unlike the symptoms of chloride toxicity, which begin at the apex of the leaves, typical sodium symptoms appear in the form of burns or necrosis along the edges (DIAS, BLANCO, 2010).

Excess sodium in the soil, more precisely, with PST (percentage of exchangeable sodium) above 15 per cent, can lead to a disastrous problem in the soil, called sodicity. In addition to causing plant toxicity, sodium can cause the soil to de-structure. Generally speaking, sodic soils, i.e. soils with an excess of exchangeable sodium, have permeability problems and any excess water will cause the soil surface to become waterlogged, preventing seed germination and plant growth due to a lack of aeration.

Results obtained by Alencar (2007) showed a low concentration of sodium, which was lower than calcium, showing that it was not a problem. Lima et al. (2007) found that there is a relationship between EC x Na^+ , with EC values (0.5, 1.0, 1.5, 2.0, 2.5 dS m^{-1}) and Na^+ (3.67, 7.49, 10.96, 13.96, 16.82 and 29.77 $mmol_c$ L^{-1}). Maia, Morais and Oliveira (1998) obtained a relationship coefficient R equal to 0.87, showing a strong relationship between these variables.

The RAS (sodium adsorption ratio) showed the highest values in wells 4 and 10 and 11, with 2.76 and 2.46 mg L^{-1} respectively (figure 32U). Similar results were found by Medeiros et al. (2003), where they observed values of 2.31 to 2.43 mg L^{-1} . Alencar (2007) observed that the wells in the locality of Gangorra showed the highest average values in the I^a and 4ª readings, which were 7.21 and 4.74 mg L^{-1} , respectively, and the locality of Baraúna, the lowest average values, which were 1.09 and 1.37 mg L , respectively.[-1]

Sodicity, determined by the sodium adsorption ratio (SAR) of the irrigation water, refers to the effect of the sodium contained in the irrigation water, which tends to increase the percentage of exchangeable sodium in the soil (PST), affecting its infiltration capacity (PIZARRO, 1985). Careful control of the water used for irrigation is of great importance, especially when it has low electrical conductivity (EC) and higher sodium adsorption ratios (SAR), which can favour the dispersion of colloids. Oliveira and Maia observed that RAS 1.6 to 3.5 mg L^{-1} in wells in the Mossoró-Baraúna region.

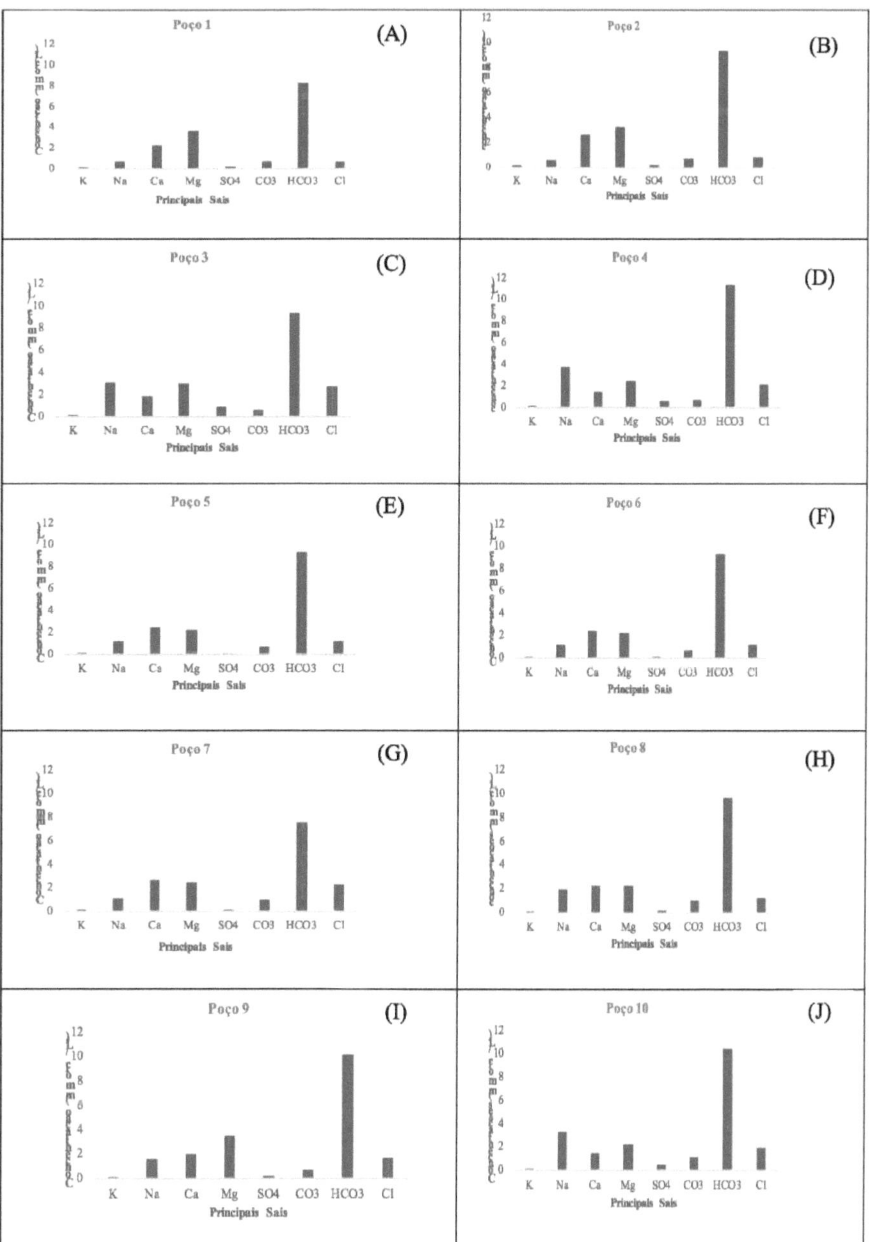

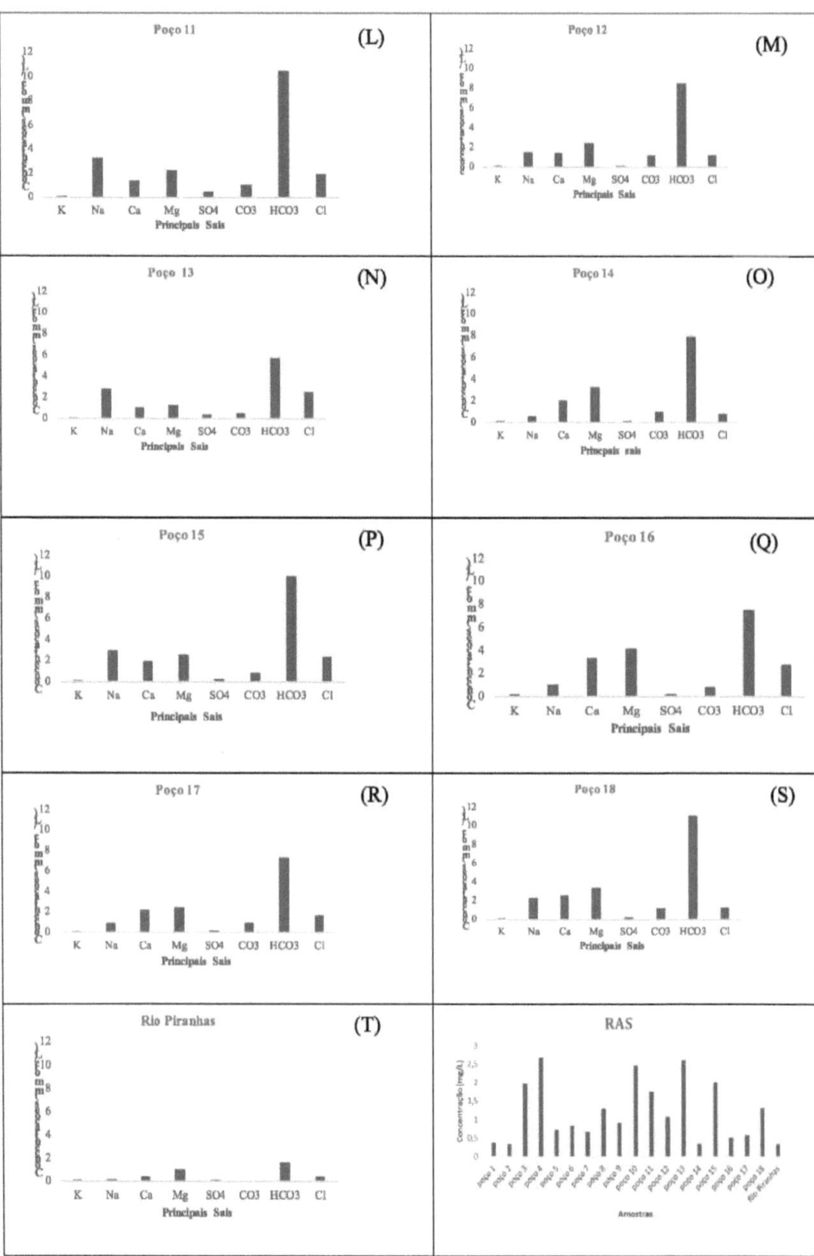

Figura 32:   Concentrations of the main salts found in the wells of the Bezerra community and várzea comprida das Oliveiras

CHAPTER 5

CONCLUSIONS

- There was a reduction in the electrical conductivity of the water from February onwards, a period marked by the start of the rains, demonstrating the effect of rainwater in diluting the salts.
- The Piranhas River showed higher turbidity values from February onwards, as the rains during this period are able to stir up the material deposited at the bottom of the river and wash away that which was in the riverbed.
- Of the two salts studied, bicarbonate was the ion with the highest values in the water, with wells 4, 6 and 10 showing 11.3, 10.34 and 10.42 $mmol_c$ L^{-1}
- Chloride and Sodium showed results varying from 0.4 to 2.7 $mmol_c$ L^{-1} and 0.2 and 3.68 mmolc L^{-1} respectively, not causing concern for use in irrigation, taking into account efficient irrigation management.
- The sodium adsorption ratio showed the highest values in wells 4 (2.67 mg L^{-1}) and 13 (2.6mgL)$^{-1}$

BIBLIOGRAPHICAL REFERENCES

ANDREOLI, C.V. Influence of Agriculture on Water Quality. Curitiba. OPS. 15 p, 1993.

AMARAL, A. B. Evaluation of underground and surface springs in the Córrego Sossego basin considering their use for domestic supply and irrigation - contamination by pesticides. Dissertation (Master's Degree in Environmental Engineering). Federal University of Espírito Santo. Vitória/Espirito santo, 2011

ALMEIDA FILHO, P. C. de. **Evaluation of the environmental and hygienic-sanitary conditions in the production of leafy vegetables in the suburban horticultural centre of Vargem Bonita, Federal District.** Dissertation (Master's in Environmental Planning and Management). Catholic University of Brasília, Brasília. 103p., 2008.

AESA: Executive Water Management Agency of the State of Paraíba, [online] available at: http://www.aesa.pb.gov.br/comites/piranhasacu/ accessed on: 10 Jan 2014

AYERS, R.; WESTCOT, D. W. **Water quality in agriculture.** 2.ed. Campina Grande: UFPB, 1999,153p. FAO. Irrigation and Drainage Studies, 29 revised 1.

AYERS, R.; WESTCOT, D. W. **Water quality in agriculture.** "Water Quality for Agriculture. FAO. Translated by Gheyi. H. R. & Medeiros, JF de, UFPB.Campina Grande- PB, 217p.1999.

BERNARDO, S.; SOARES, A. A.; MANTOVANI, E. C.. 7.ed. Viçosa: UFV - University Press, 61 Ip, 2005.

BRAZIL. Ministry of Health. *Law No. 7.802,* of 11 July 1989. Provides for research, experimentation, production, packaging and labelling, transport, storage, marketing, commercial advertising, use, import, export, final destination of waste and packaging, registration, classification, control, inspection and supervision of pesticides, their components and the like, and makes other provisions. Federal Official Gazette. Executive Branch. Brasília, DF: National Congress, 1989.

BRAZIL. Decree no. 4.074, of 4 January 2002. Regulates Law No. 7.802, of 11 July 1989, which provides for research, experimentation, production, packaging and labelling, transport, storage, marketing, commercial advertising, use, import, export, the final destination of waste and packaging, registration, classification, control, inspection and supervision of pesticides, their components and the like, and makes other provisions . Available at: <http://www.planalto.gov.br/ccivil_03/decreto/2002/D4074compilado.htm>. Accessed on: 28 June 2013.

BRAZIL. Resolution No. 357 of 17 March 2005. Provides for the classification of water bodies and environmental guidelines for their classification, as well as establishing the conditions and standards for discharging effluents, and other provisions. Available at:<http://www.unesp.br/pgr/pdf/resolucao35705conama.pdf>. Accessed on: 20 August 2013.

BRAZIL. Resolution No. 396, of 3 April 2008. Provides for the classification and environmental guidelines for the classification of groundwater and other measures. Available at: < http://www.cetesb.sp.gov.br/Solo/agua_sub/arquivos/res39608.pdf>. Accessed on: 20 August 2013.

BRAZIL. Decree no. 4.074, of 4 January 2002. Regulates Law 7.802, of 11 July 1989, which provides for research, experimentation, production, packaging and labelling, transport, storage, marketing, commercial advertising, use, import, export, the final destination of waste and packaging, registration, classification, control, inspection and supervision of pesticides, their components and the like, and other measures : <http://www.planalto.gov.br/ccivil_03/decreto/2002/D4074compilado.htm>. Accessed on: 28 August 2013.

BRAZIL. Ministry of Health. Health Surveillance Secretariat. Manual of procedures for environmental health surveillance related to the quality of water for human consumption.Brasília: Ministério da Saúde, 284 p. - (Series A. Normas e Manuais Técnicos), 2006.

BARRETO, Aurelir N. Global efficiency of water use in irrigated agriculture. In: Brazilian Water Resources Symposium. Aracaju, 2001. Proceedings... Brazilian Water Resources Association.

BRANCO, S. **M. Hidrologia aplicada à engenharia sanitária.** 3. ed. São Paulo: CETESB, 1986. 640 p.

CETESB. **Environmental and health significance of water and sediment quality variables and analytical and sampling methodologies.** São Paulo: 2009. Available at: <http://www.cetesb.sp.gov.br/agua/aguas-superficiais/125-variaveis-de-qualidade-das-aguas- e-dos-sedimentos>. Accessed on: 18 January 2013.

CARNEIRO, Joaquim Osteme. Origin and evolution of irrigation in north-eastern Brazil. In: Revista do InstitutoHistórico e Geográfico Paraibano, Ano XC, n° 34. JoãoPessoa: IHGP, 2001.

CURI, ROSIRES C. et al. Profitability of an irrigated perimeter as a function of reservoir operation and system sustainability guarantee. Eng. Agríc., Jaboticabal, v.21, n.3, p.236-246, sep. 2001.

DEUBERT, K. H. Environment fate of common peat pesticides: factors leading to leaching. **USGA**

Green Section Record.Ann Arbor, v. 28, n. 4, p. 5-8, 1990.

EUROPEAN COUNCIL. Council Directive 98/83/EC of 3 November 1998 on the quality intended for human consumption. **Official Journal of the European Communities,** 1998.L330, p. 32-54.

ESTEVES, F. A. **Fundamentos de Limnologia.** Rio de Janeiro: Interciência. 602 p, 1998.

FRANÇA, Francisco Mavignier Cavalcante (coord).Políticas e estratégias para um novo modelo de irrigaçãodocumento síntese. Fortaleza: Banco do Nordeste, American Development Bank and Ministry of National Integration, 2001.

FRAVET, A. M. M. F. **Quality of water used for irrigating vegetables in the region of Botucatu - SP.** Master's dissertation. Faculty of Veterinary Medicine and Zootechny, São Paulo State University, 2006.

FERREIRA, P. de A. **Quantification and analysis of water use in irrigated agriculture practices in the Descoberto Basin - DF.** 2005. 152 f. Dissertation (Master's Degree in Environmental Technology and Water Resources) - Faculty of Technology, University of Brasília, Brasília, 2005.

GHEYL, H. R. et al. Management and control of salinity in irrigated agriculture. João Pessoa: UFPB, 1997.

HELLER, L. et al. Third edition of the World Health Organisation guidelines: what impact to expect from Ordinance No. 518/2004? In: CONGRESSO BRASILEIRO DEENGHARIASANITÁRIA E AMBIENTAL, 23., 2005, Campo Grande, MS. **Proceedings...** Rio de Janeiro: ABES, 2005.

HELLER, L.; PÁDUA, V. L.de. **Water supply for human consumption.** BeloHorizonte: UFMG, 2006.

HU, H.; KIM, N.K. Drinking-water pollution and human health. In: CHIVIAN, E. *et al.* (Ed.). *Criticai conditiorr.* human health and the environment. 2. Ed. USA: MIT Press, 1994. p. 31- 45.

BRAZILIAN INSTITUTE OF THE ENVIRONMENT AND RENEWABLE NATURAL RESOURCES - IBAMA. Pesticides and related products commercialised in 2009 in Brazilan environmental approach. 84 p. 2010.

LARCHER, W. **Plant Ecophysiology.** Translated by C. H. B. A. Prado and A. C. Franco. São Carlos: RiMa, 2000. 53 Ip.

LIBÂNIO, M. **Fundamentals of water quality and treatment.** Ed. Átomo -Campinas - SP. 2005.

LOPES, M. E. P. de A. **Evaluation of water use in localised irrigation systems for coffee and papaya crops.** 2006. 148 f. Dissertation (Master's Degree in Environmental Engineering) Postgraduate Programme in Environmental Engineering, Federal University of Espirito Santo, Vitória, 2006.

MARASCHIN, L. **Evaluation of the degree of contamination by pesticides in the water of the main rivers that form the Pantanal of Mato Grosso.** Dissertation (Master's Degree in Health and the Environment) - Postgraduate Programme in Health and the Environment, Institute of Collective Health, Federal University of Mato Grosso, Cuiabá, 2003.

MEDEIROS, M.L.M.B.; NIEWEGLEWSKI, A.M.A; FOWLER, R.B.; ROLAND, T.R.; ZAPPIA, U.R.S.; FRANCO, P.L.P. Problemática de Agrotóxicos no Paraná. Curitiba, SURFHMA, 1988. 14

p.

MENEZES, C. T. **Method for prioritising actions to monitor the presence of pesticides in surface water:** A study in Minas Gerais. Dissertation (Master's Degree in Sanitation, Environment and Water Resources, Sanitation) - Postgraduate Programme in Sanitation, Environment and Water Resources, Federal University of Minas Gerais, Belo Horizonte, 2006.
Medeiros, J.F. de. Gheyi, H.R. Management of the soil-water-plant system in soils affected by salts. In: Gheyi, H.R; Queiroz, J.E.; Medeiros, J.F. de. Manejo e Controle da salinidade na agricultura irrigada. Campina Grande: UFPB, SBEA, 1997. Chap.8, p.239-284.

Molozzi, J.; Pinheiro, A.; Silva, M. R. da Water quality at different stages of irrigated rice development. Pesquisa Agropecuária Brasileira, v.41, n.9, p.1393-1398, 2006.

MORETTI, C. L. Good practices for vegetable production. **HorticulturaBrasileira,** v. 21, n. 2, July, 2003 - CD Supplement.

MORAES, A. J. **Manual for assessing water quality.** São Carlos: Rima, 2001. 43p.

MANTOVANI, E. C. Coffee tree irrigation. In: ZAMBOLIM, Laércio. **Coffee: productivity, quality and sustainability.** Viçosa: UFV, Department of Phytopathology, 2000. Available at : <http://www.sbicafe.ufv.br/PDF/Conteudo/155552_Art01f.PDF>. Accessed on: 29 June 2011.

MILHOME, M. A. L.; SOUSA, D. O. B.; LIMA, F. A. F.; NASCIMENTO, R. F. Avaliaçãodo potencial de contaminação de águas superficiais e subterrâneas por pesticidas aplicados naagricultura do Baixo Jaguaribe, CE. **Revista Brasileira de Engenharia Sanitária eAmbiental,** v.14, n.3, p.363-372, jul./set. 2009.

NASCIMENTO, A. R. et al. Incidence of Escherichia coli and Salmonella in lettuce *(Lactuca sativá).* **Food Hygiene,** São Paulo, v.19, n.128, p. 121 - 124, 2005.

OLIVEIRA, E. S. DE. **Geoenvironmental indicators of water quality in theCórrego Sujo basin, mid-valley of the Paraíba do Sul River.** Thesis (Doctorate in Public Health and the Environment) - Postgraduate Programme in Geosciences, Federal Fluminense University, Niterói, 2007.

OLIVEIRA, M. L. S. et al. Microbiological analysis of lettuce *(Lactuca sativaL.)* and tomato *(SolanumlycopersiconL.)* sold in street markets in the city of Belém, Pará. **Food Hygiene,** São Paulo, v.19, n.143, p.96-101,2006.
OLIVEIRA, C. A; GERMANO, P. M. Estudo da ocorrência deenteroparasitas em hortaliças comercializadas na região metropolitana deSão Paulo, SP. **Revista de Saúde Pública,** v.26, p.283-289,1992.

PASCHOAL, A.D. Pragas, Praguicidas e a Crise Ambiental: Problemas e soluções. FGV, Rio de Janeiro, 1979. 102 p.

PASCHOAL, A.D. Biocides - death in the short and long term. Rev. Bras. Tecnol. Brasília, v. 14 (1) p.24-40. 1983.

PACHECO, M. A. S. R. et al. Hygienic and sanitary conditions of vegetables sold at CEAGESP in Sorocaba - SP. **HigieneAlimentar,** São Paulo, v.16, n.101, p.50-55, 2002.

PIZARRO CABELLO, F. Riegos localizados de alta frecuencia (RLAF): goteo, microaspersion, exudacion. Madrid: Ediciones Mundi-Prensa, 47Ip. 1990.

PORTO, R.L.; BRANCO, S.M.; CLEARY, R.W.; COIMBRA, R. M.; EIGER, S.; LUCA,S.J.; NOGUEIRA, V.P.Q.; PORTO, M.F.A. **Hidrologia Ambiental.** Edusp- Editora daUSP.441p.1991

PRUSKI, F. F.; PEREIRA, S. B.; NOVAES, L. F.; SILVA, D. D.; RAMOS, M. M.Average annual precipitation and long-term average specific flow in the São Francisco Basin. **Revista Brasileira de Engenharia Agrícola e Ambiental,** Campina Grande, v.8,n.2/3, p.247-253, 2004.

RIGOBELO, E. C.; MINGATTO, F. H.; TAKAHASHI, L. S.; ÁVILA F. A. de. Physical, chemical and microbiological standards of water from rural properties in the Dracena region. **Rev.Acad., Ciênc. Agrar. Ambient.,**Curitiba, v. 7, n. 2, p. 219-224, apr./jun. 2009.

RIBAS, P. P.; MATSUMURA, A. T. S. The chemistry of pesticides: impact on health and the environment. **Revista Liberato.** v.10, n.14, p.149-158, jul./dez. 2009. Available at:<http://www.liberato.com.br/upload/arquivos/0120110910074119,pdf>. Accessed on: 01 June 2013.

RODRIGO LÓPEZ, J., HERNÁNDEZ ABREU, J. M.; PÉREZ REGALADO, A;GONZÁLEZ HERNÁNDEZ, J. F. **Localised drainage.** Madrid: Mundi-Prensa, 504p, 1992.

ROMPRÉ, A.; SERVAIS, P.; BAUDART, J.; DE-ROUBIN, M. R.; LAURENT, P. Detection and enumeration of coliforms in drinking water: current methods and emerging. **JournalofMicrobiologicalMethods,** [S.I.], v. 49, p. 31-54, 2002.

SÁ BARRETO de, F. M. **Contamination of groundwater by pesticides and nitrate in the municipality of Tianguá, Ceará.** Thesis (Doctorate in Civil Engineering, Environmental Sanitation) - Postgraduate Programme in Civil Engineering, Federal University of Ceará. Fortaleza, 2006.

SARDINHA, D. S.; CONCEIÇÃO, F. T.; SOUZA, A. D. G.; SILVEIRA, A.; DE JULIO, M.;GONÇALVES, J. C. S. I. Evaluation of water quality and self-depuration of Ribeirão doMeio, Leme (SP). **Revista Brasileira de Engenharia Sanitária e Ambiental,** v.13, n.3,p.329-338, 2008.

SANCHES, S. M.; SILVA, C. H. T. P.; CAMPOS, S. X.; VIEIRA, E. M. Pesticides and their respective risks associated with water contamination. **Revista de Ecotoxicologia e Meio Ambiente,** Curitiba, v.13, p. 53-58, Jan./Dec. 2003.

SILVA, J. G. F. da; MANTOVANI, E. C.; RAMOS, M. M. Localised irrigation. In:MIRANDA, J. H. de; PIRES, R. C. de M. (eds.). **Irrigation** (Agricultural Engineering Series). Piracicaba-SP: FUNEP, v.2, p.259-309, 2003.

SOUTO, R. A. de. **Sanitary evaluation of irrigation water and lettuce (£"c7wc"** *sativa l.)* **produced in the municipality of Lagoa** Seca, Paraíba.Dissertation (Master's Degree in Agronomy). Federal University of Paraíba, Areia. 70 p., 2005

TAKAYANAGUI, O. M. et al. Inspection of vegetables sold in the municipality of Ribeirão Preto, SP. **Revista da Sociedade Brasileira deMedicina Tropical,** Uberaba, v.34, n.l,p.37-41,2001.

TELLES, D. D. A. Water in agriculture and livestock farming. In: REBOLSAS, A. C.; BRAGA, B.;TUNDISI, J. G. (Org.). **Águas doces do Brasil: capital ecológico, uso e conservação.** 2. ed.São Paulo: Escrituras, 2002.

VON SPERLING, E. Simplified Monitoring of Surface Water Sources. In: 21°CONGRESSO BRASILEIRO DE ENGENHARIA SANITÁRIA E AMBIENTAL, 4, 2001,João Pessoa: ABES, 2001. p.1-3.

VON SPERLING, M. **Introdução à qualidade das águas e ao tratamento de esgotos.** 3. Ed. - Belo Horizonte: Department of Sanitary and Environmental Engineering; Federal University of Minas Gerais, 2005.

50

I want morebooks!

Buy your books fast and straightforward online - at one of world's fastest growing online book stores! Environmentally sound due to Print-on-Demand technologies.

Buy your books online at
www.morebooks.shop

Kaufen Sie Ihre Bücher schnell und unkompliziert online – auf einer der am schnellsten wachsenden Buchhandelsplattformen weltweit! Dank Print-On-Demand umwelt- und ressourcenschonend produziert.

Bücher schneller online kaufen
www.morebooks.shop

Printed by Books on Demand GmbH, Norderstedt / Germany